MW00459786

The Case Against Memes

By Daniel Espinola

© Copyright 2022 Daniel Espinola

ISBN 978-1-64663-825-3

All rights reserved. No part of this publication may be reproduced, stored in a retrieval system, or transmitted in any form or by any means—electronic, mechanical, photocopy, recording, or any other—except for brief quotations in printed reviews, without the prior written permission of the author.

Published by

köehlerbooks™

3705 Shore Drive
Virginia Beach, VA 23455
800–435–4811
www.koehlerbooks.com

THE
CASE
Against
MEMES

DANIEL ESPINOLA

VIRGINIA BEACH
CAPE CHARLES

TABLE OF CONTENTS

"...DANK"

THAT ONE WORD serves as a valid, if also sarcastic and ironic, response whenever one receives a tone-deaf meme. That word, appropriated form the bowels of meme culture ("culture" being a word used loosely when describing this misguided hobby), shows utter contempt for the meme that has just soiled the text message inbox of one's phone. Nearly every time, the recipient of the disdainful one-word response usually gets the hint and stays away for a moment's reprieve (though, sadly, only a moment). The word, in all of its ironic usefulness, serves as a great tool to condition the other party as to what is and what isn't an acceptable way for functioning members of society to communicate their ideas.

While memes are not an inherently new form, they have taken on a new form with the advent of the internet. The most common form of memes (and what this book is going to focus on) is the image macro meme, with two blocky lines of text. During the Great Recession, the social media pioneer MySpace had reached its peak, having already spread like wildfire through the United States. Teenagers everywhere asked each other "Do you have a MySpace?" One of MySpace's largest selling points was complete portfolio customization. The low barrier of entry encouraged nearly everybody to tailor the aesthetics of their page to suit their personalities. The site had a bulletin board app where one could post memes on their profile page, ripe for all visitors to behold. The mixture of customization and low barriers to entry served to propagate memes.

From there, memes have begun to proliferate, often through pirated means and without duly crediting the source of the content creator. The low-brow art forms started to appear on social media platforms, such as Facebook, Twitter, Instagram, Reddit, and others. What's worse is that other media outlets, the major established TV networks whom we have once relied on to deliver accurate and meaningful news, have now begun to use memes and other social media *shitposts* (a term that will be discussed later) as a credible source of news. Memes, therefore, have made a leap from MySpace to *MSNBC*.

This sort of cross-modal leap is not unlike a pathogen that crosses the species barrier. Case in point, ask anybody under the age of thirty-five what "viral" means and you're far more likely to hear a definition about something they saw on the internet, rather than nucleic acids wrapped in a protein coat. In fact, the originator of the term *meme*, famed evolutionary biologist Richard Dawkins, once stated, "when you plant a fertile meme in my mind you literally parasitize my brain"—comparing the medium to a parasite.

What's worse is that these seem to have become so commonplace in society that those with anti-meme views are often demonized for voicing an anti-meme opinion, often vehemently so. Memes have gripped this nation, and other developed nations around the world, with an almost addiction-like quality (as we'll explore later). Memes have come to the point where they are so far beyond reproach that to even question them is akin to blasphemy. However, one must ask *why* has it become taboo to be skeptical of memes? As we'll explore later in the book, we'll examine the dismissive tone that often surrounds the medium.

Speaking of blasphemous tendencies, memes have become tools of those with malicious intents to signal their radicalized beliefs to others. As we'll observe on dark corners of meme-hosting sites such as 8kun (formerly 8chan) and 4chan, memes are used to signal religious violence toward those of a different belief system (the functional definition of terrorism). However, religious zealots are not the only ones who use the flawed medium to wreak havoc—lest we forget Cambridge Analytica influence on the 2016 election. The UK-based company gamed the Facebook algorithm to steer and radicalize the American public, on a social network known for its purveyance of memes. More recently, a group of Redditors manipulated the stock market (or the "stonk" market as they preferred to call it) to inflate the price of GameStop, a company that wasn't profitable and has been fraught with strained employee relationships and poor customer service for the last several years.

This book will pick up the mantle of examining the oft-neglected pitfalls of memes, as seemingly nobody else will. As we'll explore later, Big Academia and Silicon Valley have little incentive to critique memes in earnest. Memes are dangerous, to both our communities and our society at large. Memes are also addictive, not creative, and not conducive to telling the entire story via willfully omitting context. This book will cover the negative consequences that memes inflict upon all of us.

To clarify one point in particular, this book will be unabashedly and unapologetically anti-meme. Valid peer-reviewed scientific studies and scholarly articles written by experts in various fields have been sourced and cited for the research of this book. However, it is important to keep in mind that these have been read and examined through a meme-skeptical lens. The purpose of stating this stance in the introduction is to quell any future rebuttals, in a sense, to take the power away from such accusations. After all, there is a reason this book is called *The Case Against Memes* and not *A Fair and Balanced Examination of Memes.*

The motivation for writing this book was simple; society has enough meme creators, posters, and apologists already. The meme community has been known to silence dissenting opinions. An attempt to silence opinions that one disagrees with is a bush-league tactic. The pro-meme stance cannot be a strong one if censorship is the *go-to* tactic. Worse yet, the general public seems nonchalant or outright dismissive on the issue.

However, what this book will ultimately recommend are not legislative changes—they're social ones. This book supports freedom of choice and does not advocate for any congressional actions in an attempt to *crackdown* on memes and their platforms. Given the power that Silicon Valley currently possesses over our elected officials, such a pitch would be foolish. It is also important to note that this book will not advocate for any ideology, be it political, social, religious, or otherwise. The aim of this book is to display the harm that memes cause to society and make a strong enough case that the general public will not *want* to meme anymore, even if they have the opportunity to do so. In refusing to silence memes, this book will inherently rise above the censorship tactics employed by the pro-meme community.

The format of this book will address the overarching themes that surround memes. Each chapter starts with a premise that meme-apologists often tout, with the ensuing chapter presenting the case against that premise. It is important to note that while I will discuss common meme-hosting sites (such as Facebook, Reddit, 4chan, and more), the intent is to focus on the communication medium as a whole, rather than the platform it is hosted on (though there are some instances where critiquing the platforms will be necessary due to the inherent nature of the medium).

The goal of this book is to educate the public—not to simply be a contrarian. The hope is that we create a society where ideas are not only expressed freely but are deeply pondered first. Rather than *shitposters* and pirates seeking Reddit Gold, we need scholars and polished public speakers who communicate in well-wrought sentences capable of delivering context as well as content. The hope is that, in exposing the flaws built into memes, we'll collectively abandon the medium. So, come, let us lock arms in unison, and venture into the intellectual abyss of meme-culture, for we might just resurface to a better society—a post-meme society.

BEFORE WE CAN DEFEAT meme-culture, it is helpful to define what that term is. When I refer to memes, I am referring to still images with large blocky text, often with only two lines overlaid on the top and bottom of the image. These are commonly found on Facebook feeds, subreddits, dark chat rooms, text message inboxes, and more. This definition will also include cartoon/short comic characters that are found exclusively on the internet (think Pepe the Frog or the "This is Fine" dog).

To understand the scope of the problem in front of us today, it is useful to go back in time to examine how it is we got to this point. As discussed in the introduction, the early days of the consumer internet took place in the 90s, However, that isn't really where the story begins. The turn of the millennium was the turning of the tide as more and more Americans had this great technology at their fingertips. Soon, Americans everywhere were checking bank statements, sending emails, buying goods and services, and even enjoying some alone time on adult websites, all through the power of the internet.

Sadly, the proliferation began in late 2003 with the founding of MySpace. Tom Anderson, along with his eUniverse colleagues Chris DeWolfe, Brad Greenspan, and Josh Berman founded the social media site. MySpace launched in January 2004, having been inspired by the social media site Friendster, which allowed users multiple modes of communication. In other words, Friendster allowed users to have more than just text; photos, music, and text posts were all available to users on the website. MySpace quickly expanded its user base, with one million users only a month later and five million users by November 2005. Traditional media tycoon Rupert Murdoch purchased Myspace's parent company in July 2005, as the website had become one of the top-ten most visited domains in the United States. This trend continued until the websites ultimate peak in December 2008 before beginning its continual decline.[1]

MySpace's audience skewed toward a young and tech-savvy demographic. It is unclear if Tom Andersen ("MySpace Tom") was intentionally targeting the members of society who hadn't fully developed their prefrontal cortexes, including their language, reasoning, and decision-making skills yet. One thing is clear, that is the crowd that flocked to the fledgling social media site, as it enabled users to customize their beloved profile pages. The bulletin board application that one could install onto the page enabled the proliferation of the still-relevant memes. Users could display their beloved (if not pre-selected and screened by MySpace itself) memes, in all of their tackiness and brevity.

However, as with all communication mediums, brevity comes at the sacrifice of nuance. Take away the unaesthetic blocky text

1 Stenovec, Timothy. "Myspace's Biggest Moments: Memories of a Fallen Social Network." *HuffPost*, 29 June 2011, www.huffpost.com/entry/myspace-history-timeline_n_887059?slideshow=true#gallery/5bb385dce4b0fa920b9b4ab1/18. Accessed 15 Nov. 2021.

overlaid on the screen and the viewer is left with an unassuming image. This image is not always immediately clear what it is trying to convey. The tried-and-true "a picture is worth a thousand words" doesn't hold up with memes. As one example, a classic meme format is the good advice and bad advice mallard. These two images, both of the same breed of duck with only a minor color difference between the two, can often be confused with one another. An example is posted below.[2]

This can have a dubious meaning because, on the surface, it can seem like good advice. There is also a chance that the advice found on Facebook may actually lead to positive outcome. With knowing the greater context, one could easily take this advice at face value and be lead astray.

Many memes also have dubious meanings, and solely depend on their captions to create interpretive context. Accurately grasping their meanings, absent the captions, require the viewer being up-to-speed on the culture. "Bad Luck Brian," for example, shows a photo of a seemingly awkward teenage boy. Those "in on it" know that this teen seemingly cannot catch a break, and often falls victim to bad luck and tragedy. Without the benefit of insider knowledge or

2   AriesStark. "Malicious Advice Mallard." *Imgflip*, 2015, imgflip.com/i/fzkl1.

its heavy-handed captions, the meme could easily be misconstrued as mocking the boy (one who debatably looks developmentally disabled).

Not that long ago, magazine articles, books, and long-formed essays once had an iron-clad grip among with the American reading public, with robust marketing campaigns and distribution networks. However, things have changed. Rarely does a nonfiction *New York Times* bestseller exceed more than one million copies sold. By contrast, the subreddit (a subcommunity of Reddit, denoted by the prefix "r/") r/DankMemes has a base of over five million users. We have become a society that has shunned long-form and well-researched mediums written (or at least informed) by experts. Instead, we have embraced memes with opaque origins that often grew out of the narrowest of inside jokes. Often, these memes may very well be pitching *a fraction of* one side of the story.

Academic research has unwittingly contributed to the problem, by replicating some of the same arbitrary standards as meme culture itself. Many of the academic studies cited in this book that are of a qualitative nature rely on widely known meme databases, most notably being Know Your Meme. Sara Cannizzaro a computer scientist from the University of Lincoln in the UK, aptly points out that many of these online databases lack the analytical and academic rigor associated with the term "database."[3] A database that contains little to no hard data can hardly be called a database at all. The term "collection" is more appropriate. On the other hand, quantitative studies (especially the ones referenced in this book) do not share this same pitfall.

The academic community frequently laments that memes are an understudied (and hence, underfunded) discipline for scholarly research. Yet the field of memetics (the scientific study of memes)

---

3   Cannizzaro, Sara. "Internet Memes as Internet Signs: A Semiotic View of Digital Culture." *Sign Systems Studies*, vol. 44, no. 4, 31 Dec. 2016, pp. 562–586, 10.12697/sss.2016.44.4.05

existed briefly at the turn of the millennium. A peer-reviewed publisher, *Journal of Memetics: Evolutionary Models of Information Transmission*, premiered in 1997. However, the scientific field of memetics did not last long, as it was fraught with inconsistencies. The aforementioned *Journal of Memetics* later disbanded in 2005. Several of the journal's own studies critiqued the underlying model of memetics, particularly its incorrect assumptions regarding the spread of memes. The field of memetics later was enveloped by other disciplines such as established fields of journalism, semiotics, and psychology, as Cannizzaro also discusses. So the claim by modern sociologists that memes are an understudied and underfunded research topic is not accurate; it's been tried before.

One of the downfalls of memetics as a scientific discipline was the tendency of researchers and aficionados to view memes in a vacuum rather than in their larger cultural context. As we'll explore later in this book, memes as a communication medium are not capable of creating context; rather, they feed off of it to survive. This supports that idea that memes are not a medium capable of standing independently (a core tenet of this book). Many memetic studies did not understand (or simply ignored) this fact.

Since the fall of memetics as an independent scientific discipline, researchers of memes have come to understand the importance of viewing memes as a piece of their greater context.[3] Despite the financial incentive scholars have to exaggerate the durability of memes, most have conceded that memes cannot exist in isolation— they need to be supported by a large piece of cultural norms or a longer-form communication medium. In other words, memes are nothing without context, which they are incapable of creating. This book uses meme-based studies that occur after 2005, therefore avoiding the pitfalls of poor methodology. (However, methodological flaws with meme-based studies will still be discussed.)

Even Richard Dawkins's foundational hypothesis regarding memes has been called into question. Dawkins famously asserted that memes spread by transmission and copying. Dawkins, despite being a biologist by training, neglected the concept of variability in his idea for memes.[3] Variability within a set of genetic traits enables evolution through natural selection. The same is true for internet memes. We have observed the same template countless times (for example "Bad Luck Brian") while being paired with countless text combinations. Many memetic studies cite Dawkins's work early in their introduction sections without realizing or acknowledging the flaws of his original idea. Then, these memetic studies continue to base their work off of his idea anyway. With such a shaky foundation, qualitative memetic studies were methodologically flawed from the start.

Amanda du Perez and Elaine Lombard, two visual arts researchers from the University of Pretoria in South Africa, wrote an article discussing how memes posted to one's online platforms impact their online persona, which, in turn, reveals what their offline persona really is.[4] The paper starts off by framing Richard Dawkins's original definition of a meme, though later concedes that the definition of what actually constitutes a meme is flexible and vague. This has been typical of the meme-based academic field, as massaging the definition to fit the narrative has been a go-to maneuver. This loose definition has enabled those with an agenda to fit tangentially connected phenomena into the definition of a meme. However, there are scholars who have attempted to put hard limits on the definition of the word. Amid such internal strife within the meme community, it is sometimes difficult to discern a valid study on the subject from a shoehorned one. For the purposes of this book, only quantitative studies and studies that used the word meme as a common image macro with a caption were referenced.

Dawkins alludes to many points of contention with the modern

---

4 Preez, Amanda du, and Elanie Lombard. "The Role of Memes in the Construction of Facebook Personae." Communicatio 40, no. 3 (July 3, 2014): 253–70. https://doi.org/10.1080/02500167.2014.938671.

meme. While the term *meme* as Dawkins used in 1976 predated to the internet meme in its current format (referencing instead to vague cultural ideas), many of the core underlying concepts do apply to the modern meme. Also, we cannot hold Dawkins liable to the limited understanding of memes in 1976; newer versions of *The Selfish Gene* printed in 2006 and 2016 have failed to adapt to new findings.

Dawkins dedicated an entire chapter of his 1976 book *The Selfish Gene* to discussing memes. Dawkins likened memes to genes with regards to how memes spread. However, genes, Dawkins argued, are a permanent genetic record that a skilled geneticist can trace back over long periods of time. Many subsequent quantitative studies have debunked this claim. However, memes are in fact fragile and prone to repeatedly changing as they jump from one cultural environment to the next, morphing through time. Moreover, unlike genes, memes lack traceability to their original source, since memes exist in mediums fraught with piracy.

Dawkins, however, was correct when he quipped later in that chapter, "When you plant a fertile meme in my mind you literally parasitize my brain."[5] Dawkins further discusses that memes are known to provide a "superficially plausible answer to deep thinking and troubling questions." Dawkins was more prophetic than he may have realized at the time. Think of the average subreddit; the memes posted there are ones that only the dedicated followers of that subreddit will resonate with. For memes to survive in that forum, they must conform to the community's preexisting norms. They must provide superficial answers, not conducive to conveying useful information. Dawkins himself even likened memes to a doctor's placebo, suggesting they are not meant to treat an ailment but merely to pacify the patient.

Dawkins responds to critics of his ideas later on in the chapter, accusing his opponents of "begging just as many questions" as he was. He flatly admits to using a logical fallacy in his presentation. Dawkins

---

5 Richard Dawkins. *The Selfish Gene*. Oxford, Oxford University Press, 1976.

argues in poor faith, because he assumes the pro-meme conclusion is true, and then uses that as a premise to start his argument. Dawkins could not make a fallacy-free argument to advance the medium. The worst part about his argument isn't even the logical fallacy being used (after all, begging the question is more common than people realize), but it's that his initial assumption is incorrect.

Dawkins misunderstands some key facts about memes, as he rightly points out that the longevity of any one copy of a meme is relatively unimportant. Memes have a short window of cultural relevancy and serve as little more than a snapshot. Dawkins continued to use the example of a meme that is stuck in his head and will only last for the rest of his life; however, this is unlikely. Memes do not survive for long at all, often times losing their peak relevancy in a matter of hours. With this in mind, we can think of memes as a disposable medium, one not meant for long-term use. Dawkins then exaggerated the ability of memes to proliferate, keeping with his comparison to genes. He further mused that some will even have brilliant short-term successes in the meme pool (as he refers to it) but will not last over the long term. Dawkins is only half-correct in this statement. While it is true that some memes will have better success than others in their proliferation, memes do not have long-term viability. Statistically, memes lose relevancy over the span of hours, *not* decades.

Dawkins's comparisons of memes to genes has other limits as well. He suggested that copy fidelity (that is, how true to the original a copy is) is an important characteristic of survival for memes. Here, Dawkins admitted to being on "shaky ground." His doubts have been proven correct—the medium is fraught with piracy and constant creative interference from manipulative users. Often, a meme will need to be changed as it is moved from one cultural backdrop to the next. Memes are often copied and pasted without direct citation that credits the original author. Interestingly so, Dawkins himself conceded that memes are not a highly copy-fidelity medium.

Dawkins further muses the core idea that a meme is essential to "all brains that understand the theory," which underscores one of the major flaws of the medium. Memes assume that the audience will understand the content within. Especially so since memes are not good vehicles for building understanding or building cultural bridges. However, this is a mighty assumption to make when selecting a communication medium, believing that the audience will just inherently understand the message.

Dawkins continued his chapter by discussing that memes reinforce each other. Memes are often lauded for their ability to start conversations. However, the conversations that are created are merely an echo chamber (which, to be fair, *is* reinforcement, albeit a bit misleading). Dawkins also dovetails the reinforcement idea into his notion that memes survive due to deep psychological impact. However, we must ask ourselves as a society if we really want to continue to use a communication medium that relies so heavily on shock value and manipulation.

Dawkins explained memes carrying falsehoods. As Dawkins himself puts it, "Nothing is more lethal for certain kinds of meme than a tendency to look for evidence." If the great champion of the communication medium himself has debunked the medium as being plagued with lies, then we, as a society, should discontinue use of the medium. Dawkins also demonstrated the addictive quality of memes as he mused "the success of a meme depends critically on how much time people spend in actively transmitting it to other people." Memes, by their very nature, are a medium that need to be constantly propped up in order to proliferate to others. Worse yet, they distract and detract from real-world endeavors, as we'll discussion in the activism chapter.

Dawkins later talks about memes ultimately vying for attention, stating that it is hard for new memes to invade the meme pool. *Invade* is an interesting word choice by Dawkins, as it indicates the level of force necessary for survival. Memes are not subtle, nor do they ask permission. Memes as a medium are often forced upon bystanders, whether scrolling Facebook or persisting in the form of an unwanted text message.

Dawkins ends his chapter on memes likening them to genes, stating that selfish memes have no foresight. This is correct, as memes rely so heavily on their greater sociocultural environments that birthed them, often failing to stay relevant a mere hours after posting. Memes are a backward-facing medium, unable to react to the present day, instead only serving to recount past events. Other communication mediums, such as books, for example, often relive second lives, whether being formed into a movie or television series or public lecture. Memes, however, die an often-unceremonious death with no afterlife. Memes truly are disposable.

IT IS COMMONPLACE to view the production of memes as a creative process akin to the more formal work of graphic design. Yet the comparison quickly breaks down. A skilled graphic designer can create standalone images that are capable of conveying a strong, immediately clear message. There are a lot of things a graphic designer would consider when crafting an image. Color choice is very important to the skilled graphic designer, as nearly every color, including varying shades of darkness and lightness, convey a psychological subtle message. Graphic designers have long worked hand in hand with marketers to assemble the perfect branding image for a company or product. With the stakes being so high, the room for error is often razor-thin.

Font choice is yet another decision in the graphic design community that is made with due consideration. The choice of font—if words are even necessary, as a skilled graphic designer is capable of producing work that can stand alone—can significantly alter the way the viewer interprets the message. Much internet hatred has been doled out in the direction of the Comic Sans font for often cheapening a message. This was played up for dramatic effect in the 2015 PC and console video game *Undertale*, in which a vital character, Sans, has his text dialog appear only in the ridiculed font of his namesake. The player is often left to not take him seriously until the very end of the game. Other fonts, such as Time New Roman, are often used to convey a serious, albeit stiff, tone with the viewer. The point is elementary but profound—*font choice matters.*

In contrast to such technical work, *memeing* is an amateur sport, one meant for novices and others who don't fully comprehend what they are doing. However, the common meme seems to have one default text setting—a call to rally others for a cause, offering political commentary, or to humorously entertain. The following two memes below will illustrate this point.

This meme,[1] depicts a dog with a wide-open grin on his face. The dog's eyes are closed, to depict something is humorous, as if the dog itself is laughing. The text lines are clear enough to convey that this is a meme intended for humor. Here, we see that the standard

1   "Funny Animal Memes." Slapwank.com, 2017, slapwank.com/wp-content/uploads/2017/03/Funny-Animal-Memes-3.jpg. Accessed 2021.

meme text is used—large, blocky letters, white in color. The meme text could have been written in a font choice that better serves its comedic value, though the creator who first visited quickmeme.com

must not have thought of that. The next meme will depict something a little more serious.

This meme depicts famed actor Leonardo DiCaprio proposing a toast to readers who follow political news but manage to remain civil[2]. This meme is intending to convey a message of sincerity. The no-frills font fits appropriately for the message being conveyed. However, the suitability of the font is likely pure happenstance, as this is the standard default text that nearly every meme generator uses. It's also the same as the dog meme. This font selection, as it is the default setting, lumps this serious meme into the same pool of the meme-verse. That is to say, lack of attention to detail regarding this font choice has cheapened the second meme, grouping it in with its less genuine siblings. In truth, most memes eschew seriousness

2 Giovanni. "The Best 2016 Political Memes." Urban Myths, March 5, 2016. http://www.urbanmyths.com/urban-myths/politics/the-best-2016-political-memes/

in favor of heavy-handed humor. Full disclosure—finding a serious meme was shockingly difficult.

Text placement and contrast are another important decision a graphic designer makes when considering a piece of work for a client. Where the text lands on the image guides the viewers eyes, as if the artist is luring the viewer to the stronger points of the message being conveyed. A skilled designer can make words seemingly pop from the page, pull the reader through the page, or navigate them to see things in a sequential order. Font choice and word placement can also be used to "hide messages in plain sight from squares like Dan," as an old graphic design friend of mine once cheekily said. The common meme takes nothing of this sort into consideration. There is often no variation in the placement of the text, just a line on the top third of the image and the same at the footer of the image. There is no masterful guidance of the viewers eyes around the page. There is only an optical obscenity, delivered with all of the subtlety of a sledgehammer.

Another artistic concept that is often neglected within the *memeing* community is contrast. When used as a noun, contrast is the degree of difference between the lightest and the darkest part of a picture. Simply put, contrast is what makes certain elements of an image appear different from one another. Pictured below is a common meme that demonstrates the lack of attention the meme-o-sphere pays to a common element of design theory.

This meme, known as the *Overly Attached Girlfriend*,[3] perfectly shows the lack of attention to detail shown in regards to contrast. The top left-hand corner is immensely white, so much so that is starts to wash out the rest of the top left quadrant. The top sentence seemingly blends in with the lights, making it difficult to read for some viewers. The fact that the woman in the image is also standing in front of a white wall doesn't help. The bottom left quadrant of the photo also suffers from the same issue; the light is so intense that it also starts to wash out the beginning of the second phrase as well (though to a lesser degree from the first). A better choice would have been to color the font something other than white. White font overlaying a white background which is washed out by white light was a poor choice. Clearly, neither the original creator of the meme nor any of its subsequent posters paid any attention to basic graphic design concepts.

The digital meme is largely an amateur sport, with little to no barrier to entry. Type "How to make a meme" into your preferred search engine, and you'll be inundated with results for online meme

---

3   PhantasmGear. "Pin by Phantasmgear.com on Overwatch | Overly Attached Girlfriend, Real Estate Memes, Girlfriend Meme." *Pinterest*, www.pinterest.com/pin/845691636250276337/. Accessed 21 Mar. 2022.

generators. These simple programs allow a user to upload an image and type their desired text into the software. Creating a meme is as simple as a few clicks and keystrokes. Creators can now sidestep the traditional channels of mass media and art distribution to proliferate their "work" (a term I'm being loose with) to the public.

These meme-generators are not overseen by a panel of experts prior to their release. There is no constructive feedback given to their creators, nor is there a wise art-school professor giving a grade to a hopeful student. There is no careful consideration regarding the color selection of the image, its relation between the background and foreground, or the selection of its font. These are all criteria that a trained graphic designer would consider when creating a piece of media. With memes, none of these things are being considered. The resulting meme can often be visually jarring or just plain confusing to the viewer, who is absent of context. The barrier of entry as to who gets to create a meme is not the only thing that has been reduced; standards have dropped as well.

What's even worse is that this lowering of standards seems to not only be tacitly accepted by the public as a whole, but it is seemingly celebrated by the online *memeing* community. Take the term *shitposting*, for example. *Shitposting* refers to the act of creating an online post (whether it would be a social media entry, meme, or short blog entry) that creates either a stir among its naysayers or an applause among its supporters (with the least amount of effort). The less effort and more controversy, the better the *shitpost*. These posts are often a high-volume, low effort endeavor that rewards not creativity, tact, and deep thinking but a rather juvenile intellectual mischief.

In the traditional art world, by contrast, painters, sculptors, and other artists of various mediums try to create a masterpiece that fosters deeper thinking, with the hopes of sparking a grander societal discussion, in which constructive dialogue is had. Creating an aesthetically pleasing piece of artwork is also high among their priorities.

The image below is an example of how artistic standards have fallen in the meme world. On August 26, 2021, this image was found on r/DankMemes (a common hangout forum for meme aficionados), perfectly conveying the low standards of artistic brilliance in the medium.

This image looks as if it were drawn by an amateur using Microsoft Paint. The bottom two panels and the top-left panel are jagged and rough. The eyes are just two black dots with no detail. Similarly, the rest of the facial features also lack any semblance of human-like detail. The text in the top left panel is so poorly placed that the bottom sentence overlaps with the speaker's head. The bottom-right panel is the first one of these panels to have eyebrows, as if they just mysteriously appeared only after the anti-vaxxer's poor logic has been pointed out. The panel on the top right depicts a speaker who lacks any color at all, seemingly as if he were copied and pasted there on the blue background. Furthermore, the color is splashing out of the character, spraying haphazardly onto the background. The speaker's head is so poorly drawn, a portion of his head jarring upwards at a slight angle, the same way one would circle an option on a form they were filling out. Like the other three panels, the speakers face lacks any semblance of actual detail. The speaker's torso, much like their head, also suffers from a stray line, caused by lackadaisical hand drawing.

Such rock-bottom artistic standards contradict established norms. In 1994, the National Arts Education Association (NAEA) developed a list of educational standards that can be adopted by each individual state for use in public kindergarten through twelfth grade education. These standards detail a wide variety of what a student should be capable of in each successive grade level. The standards were updated in 2014.[4]

On several fronts, *memeing* falls below the level of what the NAEA states students who have finished the eighth grade should be capable of. The NAEA supports the concept of "balancing experimentation and safety, freedom and responsibility while developing and creating artworks"; the NAEA recommends that, upon completion of the eighth grade, students be able to "demonstrate awareness of practices, issues, and ethics of appropriation, fair use, copyright, open source, and creative commons as they apply to creating works of art and design." As discussed further in the book, memes as an artistic medium struggle with the concept of copyright protection and ethics as they relate to the use of another artist's work. The medium is fraught with piracy as many purveyors of memes exist solely in a copy-and-paste culture that does not lend itself to crediting authors for their work.

Another NAEA standard for eighth graders is to "create and interact with objects, places, and design that define, shape, enhance, and empower." Students are expected to be able to "select, organize, and design images and words to make visually clear and compelling

---

4   National Coalition of Core Arts Standards. *National Core Arts Standards-Visual Arts at a Glance*. Dover, Delaware, National Core Arts Standards, 2014.

presentations." This intellectual standard runs counter to the common meme; memes are stripped of their wider contexts and rely on their viewer being "in on it." That is to say, memes on their own right are not capable of clear presentations, and they often need support from a secondary source of context, such as a blog post or a Twitter feed.

An additional NAEA standard revolves around visual imagery and how it influences understanding and responses to the world around us. The standard expects eighth graders to be capable of analyzing how the viewers engage with images that influence ideals and the impact of context. The aforementioned act of *shitposting* rejects the practice of reflective engagement, as it creates low-effort posts that serve only to broadcast a message (often insincere) without concern for the viewer or the wider cultural contexts. *Memeing* as a hobby is antithetical to these standards.

However, there may be more sinister motives behind the artistic choices displayed in memes. The obnoxiously large white blocky text with the contrasting black outline isn't chosen at random. That aesthetic eyesore is chosen because it is highly visible, and hence more likely to be convincing. A study performed at the University of Michigan revealed that highly visible phrases were deemed to be more truthful than moderately visible phrases, regardless of whether or not the statement was actually true.[5]

The addition of a photo has a similar effect. A series of experiments demonstrated that when statements were accompanied with a photo, viewers were more likely to believe that the statement was true, even if it wasn't[6].

These two studies have clear implications as to the basic building

---

5   Reber, Rolf, and Norbert Schwarz. "Effects of Perceptual Fluency on Judgments of Truth." *Consciousness and Cognition*, vol. 8, no. 3, Sept. 1999, pp. 338–342, 10.1006/ccog.1999.0386.

6   Newman, Eryn J., et al. "Nonprobative Photographs (or Words) Inflate Truthiness." *Psychonomic Bulletin & Review*, vol. 19, no. 5, 7 Aug. 2012, pp. 969–974, 10.3758/s13423-012-0292-0.

blocks of memes. Large blocky text makes the phrase difficult to ignore. The color scheme makes it jarringly obvious from nearly any background. A photo to accompany the artistically challenged font all culminate in the viewer subconsciously lowering their defenses when it comes to what is true. While it isn't likely that these two studies are well known in the meme community, we would be naive from a social dogma or political radicalization point of view to believe that meme creators aren't trying to manipulate the viewer with the least possible effort.

By embracing memes, we have accepted artistic mediocrity as a society. It doesn't have to be like this—we can start by demanding better quality work from creators. It was once conventional wisdom that artists had to hone their craft for many years and become proficient in the techniques of their medium. The modern meme scene has reduced the barrier of entry to such a degree that anybody, regardless of actual skill, can call themselves a creator.

Meme hosting sites routinely neglect the bedrock of creative endeavors: copyright law (disclaimer: this book does not substitute for or offer legal advice; contact a licensed attorney for legal advice). While memes by their very nature are derivatives of some other creative work, they often give no credit of the original creator, nor do they seek permission. As such, images and other such works not in the public domain can potentially be held liable for copyright infringement. A common rebuttal from *meme apologists* is the doctrine of fair use that exists with the Copyright Act of 1976. However, a work must pass a four-point examination in order for the fair use exemption to apply. There is no blanket formula that protects all memes from infringement claims.[7]

Even when fair use does apply, it is also not a guarantee that a poster will not be sued by the lawful holder of the copyright. It is merely a defense that can be made once they are in front of a

---

7   Patel, Ronak. "First World Problems:' a Fair Use Analysis of Internet Memes." *Escholarship.org*, 2013, escholarship.org/uc/item/96h003jt.

judge. One common tenet of fair use is to examine the quantitative and qualitative aspects of the pilfered work—how much, and how important is it? The common meme, in its macro image format, relies on using a still image. The appropriation of photographs and other still images severely weaken the case of a fair use defense.[7] Another question to keep in mind with the doctrine of fair use is whether the nature of the work infringed upon was artistic or not. Courts have been less likely to uphold fair use standards for the infringement of artistic works, which happens to be the content that is pilfered most by meme creators.

Ronak Patek, a legal scholar from the University of California, Los Angeles, examined the doctrine of fair use as it relates to internet memes.[7] Patek touched upon key concepts in the doctrine of fair use—focusing on social desirability and cultural interchange. The concept of interchange implies that two or more sides exchanged something (even abstract notions, like ideas) with each other. Memes serve mostly to broadcast to a preselected group of fellow users, conforming to the norms of the group. In this, it is a stretch to say that memes promote *exchange*. Given the radicalization of contemporary United States (both politically and socially) over the past decade, reinforcing preexisting echo chambers should not be considered a *socially desirable* outcome either.

Another anchor of the fair use doctrine is the concept of public benefit. The public benefit defense is most aptly suited to things like informing the public of socially significant events. Memes are poorly suited for this purpose, so they thrive within contained or private groups. One must also ask (even if rhetorically so), how much does the public really benefit from the proliferation of memes?

Fundamentally, memes are derivative, *not* creative. In a thesis defense, Heidi Huntington, an internet researcher from Colorado State University, stated that memes serve as a means of visual rhetoric, and she explores the many ways that they do so[8]. Huntington states that memes often rely on cut-and-paste jokes and that these jokes often rely on parodying, mimicking, and recycling pieces of culture. Memes are often known for their unoriginal content and the cyclical nature of their humor. Memes are not a creative form of art; they are leftover bits of culture that are constructed with yesterday's garbage.

One simple way to detect a meme's journey through the creditless internet is the reverse image search. To perform the search, right-click over an image online and copy the image URL. Then open your preferred search engine (most of which have this option available) and navigate to the images section. Then copy the image URL or image and click search. The results that appear will show where else on the wide ocean of the internet this image has appeared. Unsurprisingly, a lot of these memes are not original content. Perform a reverse image search of the common meme on 4chan or Reddit and you will find that many of these "original" posts have appeared elsewhere before. The hypocrisy of Reddit memes are particularly glaring, since many of the popular *memeing* subreddits grandstand on their requirement to post original content only.

The ever-controversial 4chan litters new visitors with several legal disclaimers and NSFW (not safe for work) tabs. Further investigating the rules of the site, however, one discovers the rule of not attaching signatures to their posts. This provision, coupled with the fact that one does not even need a personal account to post on 4chan, shows that the website is not committed to protecting the intellectual property rights of content creators. Memes as a communication medium are riddled with plagiarism, and websites such as 4chan enable the pilfering.

Francisco Yus, a professor of English at the University of Alicante in Spain, examined the relationship between meme usage at all stages, from original examination through their spread, and he studied their

---

8   Huntington, Hiedi. "Subversive Memes: Internet Memes as a Form of Visual Rhetoric." *Selected Papers of Internet Research*, 2013. *Colorado State University.*

impact on the sense of identity of the individuals who spread the meme[9].For context, Yus discussed the difference between memes and other pieces of viral content. Viral content often spreads in an unaltered state, thus staying true to its original intention. However, most memes change as they spread from one group to the next, not unlike what happens in the children's telephone game. This results in the common meme being far removed from its original communicative purpose as it recirculates.

But what if the redditor was the original poster of that meme on the internet? the meme defender will inevitably ask. The fact that the meme has seeped into every nook and cranny of the internet only further reinforces the point of the medium being doused in copy infidelity (that is, copies are *not* true to the original). Facebook, Reddit, 4chan, and Instagram are not in the business of paying royalties to original content creators, unlike other traditional mediums of published communication. As an example, the image below was found on r/DankMemes on August 26, 2021. Performing a reverse image search showed that this very same image showed up in various corners of the internet, dating as far back as 2018, displayed on businessinsiider.com. Unsurprisingly, it was certainly not used to make a morbid joke about bombing Afghanistan. So not only are *meme-lords* stealing from one another, they are also stealing from legitimate journalists.

National bird of Afghanistan

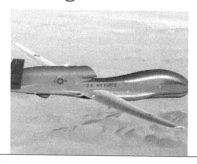

---

9   Yus, Francisco. "Identity-Related Issues in Meme Communication." *Internet Pragmatics*, vol. 1, no. 1, 28 May 2018, pp. 113–133, 10.1075/ip.00006.yus.

Even if memes and their enabling websites *did* pay royalties, how would these sites know who to send the payments to? Commonly posted memes are so overwhelmed with piracy that—since the same image appears in countless corners of the internet—determining who the legitimate source of an image is can be extremely challenging.

The word *meme* itself suggests a lack of originality. Another common pitfall discovered whilst performing research was that very few papers focus on the etymology of the word *meme* itself. Cannizzaro delves into this. The word combines the English terms "mime" or "mimicry" with the French word *même,* which translates to *same* in English. The very origin of the word denotes the weaknesses of memes as an artistic medium—mimicking other works.

Cannizzaro appropriately discusses plagiarism within the medium of memes. The paper states that the journey a meme takes over its lifespan (however short that may be) takes priority over the origin of the meme. Such disregard for citing a source properly or tracing appropriate lineages isn't evident in books, magazine articles, or scholarly articles. Accusations of plagiarism can and often does ruin careers in these fields (as they should). Cannizzaro also points out that memes cannot be traced back to their original source easily (or perhaps at all). This is because, during the spread of memes (often referred to by others as the transmission process), users will change and morph these memes before posting them. The rampant plagiarism and the refusal of users to sign or otherwise authenticate their handiwork leads to the loss of traceability within the medium.

Amanda du Perez and Elaine Lombard also discussed fidelity as one of the characteristics of a viable meme. This notion of fidelity within memes was first introduced by Richard Dawkins in his 1976 book *The Selfish Gene.* As this chapter has explored, memes as a medium are anything but true to their original source or their original intent. Memes undergo a substantial change along their journey through differing subreddits and Facebook groups. Memes are also frequently plagiarized, and the original creator is seldom credited.

Using this as a characteristic for what constitutes a viable meme, therefore, would dramatically cut down the number of memes.

Sadly, some within academia have tacitly accepted meme culture's disregard for source crediting. Chen Kertcher and Ornat Turin, two Israeli-based peacekeeping researchers, discuss the use of memes in Israel during the COVID-19 pandemic[10]. They both erroneously muse that memes as a medium "are free from legal, economic, and ethical constraints." This is indicative of how many meme users view the medium as if it were the Wild West, where societal bedrocks such as copyright law and honest communication do not apply. This is bad enough when Redditors spout this nonsense, but it is more troubling when two published academics spew attitudes like this. Statements that condone the rampant wrongdoing that is so common in the medium should not be tolerated. Just because enforcement in the medium is lax doesn't mean the rules don't exist.

---

10   Kertcher, Chen, and Ornat Turin. "'Siege Mentality' Reaction to the Pandemic: Israeli Memes during Covid-19." *Postdigital Science and Education*, vol. 2, no. 3, 5 Aug. 2020, pp. 581–587,

DO MEMES CREATE THEIR OWN CONTEXT?

WHILE MANY MEMES are intended only for entertainment, others are not so benign. One notorious example is Pepe, a green cartoon frog used by its malevolent (and often anonymous) users to tell stories of sociopathic and often extremist activities. As with Bad Luck Brian, Pepe's base form is not immediately clear what message he conveys. Pepe the Frog is a much less obvious example of white nationalism than, say, the Confederate flag. The meme relies on the context of its captions and the reader's previous experience with the meme. Pepe is a particularly heinous example of both needing to be "in on it" and of how memes are often taken advantage of by nefarious players on the internet.

But memes are meant to be consumed with their heavy-handed captions, a meme defender will inevitably say. When a meme does come with context, it is usually the thinnest serving possible. During election campaigns here in the United States, political radicals of both parties will resort to *memeing* as a means of signaling to their cohorts. Often, these memes will include a sentence at the top of the image and a sentence at the bottom of the image. The forced brevity of the medium requires the author (often anonymous or uncredited) to leave out a lot of key information. The memes must be snappy, as it needs to stand out and grab viewers' attention. This inherent need for a hook and brevity (after all, captions cannot cover up the image!) forces—or should we say encourages?—the creator of the meme to omit context. This lack of context is often intentional, bolstering the viewpoints of their party while ignoring or downplaying the views of the other side. They are not trying to win converts, rather energize the already converted. In no small way, memes have played a hand in radicalizing and dividing America. The meme below illustrates this phenomenon.

The meme portrays Joe Biden, the current President of the United States.[1] The top sentence offers a very brief synopsis of his political career. The bottom sentence makes a claim about his current job performance. The top sentence references Biden's six terms he spent as a senator from the state of Delaware. However, due to the forced brevity of the medium, it makes no references to key stances that Biden took as senator nor key pieces of legislation he wrote or voted

---

1 "Memes." *Anti-Democrat Stickers*, antidemstickers.com/memes/. Accessed 2021.

on during his tenure. The brevity of the meme forces (or better yet, entices) the creator to omit these key pieces of information.

The second part of the top line focuses on his time spent as Vice President. There were no ties in the Senate that required a tie-breaking vote during Biden's tenure as Vice President. It also neglects to mention that Barack Obama never relinquished his Presidential authorities to Joe Biden during his tenure. In other words, it fails to mention Joe Biden's relative lack of political power during his tenure as Vice President. Such a statement is irrelevant at best and a straw man (intentionally misrepresenting your opponent's argument as weak) at worst.

The bottom text line makes a claim of Biden blaming his predecessor for the country's problems. However, it makes no specific claims as to what those problems are. Nor, for that matter, does it make any citations as to any speeches or press releases where Biden has explicitly blamed the previous US President. It simply assumes that President Biden is preoccupied with complaining.

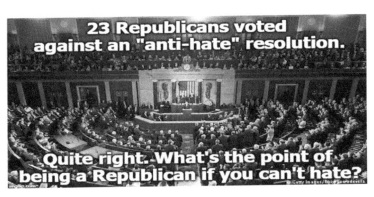

It is not the aim to play partisan politics in this book, for the United States already has enough division and radicalization. Therefore, we will examine briefly an anti-Republican meme.

Here, we are shown an image of Congress,[2] with the same tired

2 "Image Tagged in Republican,Hate,Congress." *Imgflip*, 2020, imgflip.com/ i/2vhv1w. Accessed Aug. 2021.

old format of the two blocky lines of text strewn across an image that cannot stand alone. The top line makes a claim of twenty-three Republicans voting against an anti-hate resolution. It does not name the Republicans, leaving the reader with no way of verifying that they are indeed Republicans or how they actually voted. It also leaves out the name of this piece of legislation, instead opting to say "anti-hate resolution." Without knowing that key piece of context, how can the reader of this meme know what this proposed law was called, or what it contained? As displayed on the meme, it couldn't possibly be ascertained the reasons the twenty-three alleged Republicans voted against this nameless bill.

The bottom line of blocky text asks "What's the point of being a Republican if you can't hate?" This boils the opposing party down to a group that does nothing but hate, thereby pushing a nonconstructive "good versus evil" narrative, which further radicalizes Americans. Again, without the supporting piece of information mentioned in the previous paragraph, it's difficult to know whether or not the piece of legislation had anything to do with "hate" at all, let alone who was on the receiving end of this alleged hate.

While the brevity of memes in a vacuum is concerning, we must keep in mind that memes are meant to exist in a vacuum. Memes are posted often without citing any sort of supporting evidence. It is not as if memes come with a works cited or a bibliography section. The readers are left to accept that the author performed their due diligence and isn't trying to steer them in a particular ideological direction. Readers, then, are likely to encounter the meme and trust the post, assuming it is ideologically friendly territory.

Popular meme-hosting website 4chan has a lengthy list of rules that its anonymous users supposedly must follow (although, how do you ban someone that was never a member, particularly in a tech-savvy forum where VPN use and other trace-covering tactics are common?). Among one of these rules is to not embed images posted to the site with other modes of communication, such as sound files, text documents, and archives. This forces users to omit several layers of context, often leaving the viewer with nothing more than an altered image to judge, which, as we have discussed, can paint a biased, misleading, or incomplete picture.

Intentionally misrepresenting your ideological opponent's argument is nothing new. Humans have been perpetuating this for thousands of years. It's called the straw man fallacy. For the layman, a straw man fallacy is when person A intentionally misrepresents the argument of person B by reducing it down to a very basic form or often making the argument appear weak, and then person A debunks the artificially weakened argument. Memes, in all of their aforementioned brevity, encourage one side to use the straw man fallacy on their critics.

However, with memes, the blocky captions typically only leave room for two lines, both controlled by the author. Authors, in many subreddits, for example, may post without a moderator approval. Moderators, when they are present, serve as gatekeepers. However, instead of gatekeeping for factual accuracy, they gatekeep for conformance to ideology. Thus, they often have the power to censor and silence arguments they don't like, leading to the creation of echo chambers.

Now, it is not my position to say that all memes are guilty of intentionally creating echo chambers to distort public discourse. It is, however, this book's position that all memes are *susceptible* to this pitfall, as it is an inherent danger of memes as a medium. Their forced brevity comes at the cost of both context and nuance.

Yus delves into the context of the common meme being dependent on its environment. The example of a meme revolving around butter is used in the Yus paper, signifying different things in a different background (a Facebook culinary group versus a fat-shaming board on 4chan, for example). This shows yet another flaw of the communication medium, the malleability of the common meme. A long-form text post or a book, for example, will ideally build plenty of context (assuming they are well written) and therefore can stand on their own two proverbial legs, without needing the platform itself to give meaning to the work. The common meme is weak in this aspect, as it relies on the meme-hosting site to survive and spread, not unlike a parasite.

Athina Karatozgianni, Galina Miazhevich, and Anastasia Dennisova, three media researchers at the University of Leicester in the UK, discuss the role of memes in Russian-language civil discourse.[3] Interviewees for their study proclaim that memes can convey the tone of a political event "much sharper than ten highly intellectual newspaper columns." This is such an erroneous claim to make, given the short quippy nature of internet memes. Memes rely on the viewer inherently understanding the content and the context that is offered. A skilled wordsmith, on the other hand, can create understanding in the reader where none previously existed, a trait that the common meme is incapable of.

Cannizzaro discusses that memes cannot be used to transmit ideas from one user to another and the inherent variability that such interpretations introduce into a communication medium. Since such variability exists within the medium and the medium is subject to constant alterations between users (and especially between subcultural groups such as differing subreddits, for example), we can deduce that memes cannot reliably be used to broadcast objective information.

Amanda du Perez and Elaine Lombard discussed that memes are spread if they are believed. However, memes are especially susceptible to the illusory truth effect, which basically states that if a false claim is repeatedly told, it will eventually be perceived as true. Therefore, it is entirely possible that memes are spread even, or especially, when they contain outright lies. Since memes do not come

---

3   Karatzogianni, Athina, et al. "A Comparative Cyberconflict Analysis of Digital Activism across Post-Soviet Countries." *Comparative Sociology*, vol. 16, no. 1, 13 Feb. 2017, pp. 102–126, 10.1163/15691330-12341415.

with any works cited page, there is often no way for the viewer to know the truth beyond being hyper-diligent and independently fact-checking information. (Sadly, that is not as common as it should be.)

Since the communication medium ultimately is at the mercy of its environment, and cannot create context on its own, the poster is often left unknowingly at this mercy. Meme usage, therefore, inherently signals a forfeiture of interpretation control to the masses as Amanda du Perez and Elaine Lombard noted.

DO MEMES CREATE ENGAGING CIVIL DISCOURSE?

POLITICS IS NOT THE ONLY REALM of society that have been radicalized in no small part to memes; race relations have been impacted by the choppy medium as well. In 2005, an artist named Matt Furie created a green humanoid frog by the name of Pepe. Pepe began as a completely apolitical and non-bigoted character, spreading good will to the other characters create by Furie. Thanks to the insufferable *copy-paste now, credit never* culture of common meme-hosting sites, the presence of Pepe seeped into all corners of the internet. Pepe's presence was co-opted by the small but increasingly prominent alt-right movement in 2015. The various manifestations of Pepe became commonly associated with anti-Semitic statements and have repeatedly been used to condone white nationalism.[1]

The rampant plagiarism of the common meme lead Pepe's

---

1 Swinyard, Holly. "Pepe the Frog Creator Wins $15,000 Settlement against Infowars." *The Guardian*, The Guardian, 13 June 2019, www.theguardian.com/ books/2019/jun/13/pepe-the-frog-creator-wins-15000-settlement-against-infowars.

creator to sue the right-wing website *Info Wars* for the amount of $15,000. Furie became tired of Pepe's gross radicalization and has sued in hopes of preventing Pepe's use to condone wanton hatred.

Sadly, Furie's efforts have not stopped Pepe's association with hate. In 2016, The Anti-Defamation League included Pepe the Frog in its database of hate symbols, thereby shining a light on its proliferation throughout the internet.[2] With this association now enshrined, one must now suspect something more when encountering this frog on the internet. We're not suggesting that memes caused the first radicalized activist; however, we are suggesting that memes as a medium act as fuel to the proverbial fire. It isn't clear what would have happened to the modern white nationalist movement if Pepe had not existed. However, given that symbols and logos have commonly served to unite humans around a common cause, it stands to reason that this hateful movement would be much more fractured and weaker without the meme (at least in the digital space).

The piracy of memes also helps bolster their radicalization effect on the viewer. A study concluded that repetition of a piece of information can make the statement appear true, regardless of whether or not it is.[3] This finding displays what psychologists call the illusory truth effect, which states that a person is more likely to believe a statement they've heard multiple times. Often times, this is because the listener hears the statement, and then compares it to their currently existing body of knowledge, subject to all the biases therein. This is a psychological loophole that is ripe for exploration in the medium, as those with an agenda can use memes to spout falsehoods and have the general public believe it, so long as they

2   Anti-Defamation League. "Pepe the Frog." *Anti-Defamation League*, 2016, www. adl.org/education/references/hate-symbols/pepe-the-frog.

3   Begg, Ian Maynard, et al. "Dissociation of Processes in Belief: Source Recollection, Statement Familiarity, and the Illusion of Truth." *Journal of Experimental Psychology: General*, vol. 121, no. 4, 1992, pp. 446–458, 10.1037/0096-3445.121.4.446.

repeat themselves often enough. There is no wonder, given all of this, that memes are such a purveyor of misinformation. Such findings dovetail with the copy-paste culture of common meme-hosting websites. This much dishonesty can only harm the informed societal discourse that we all strive for.

If one hallmark of civil discourse is the cogent transmission of ideas or beliefs from one person to another, then memes fail gravely on that front. The purported ability of memes to facilitate user-to-user communication has long been hypothesized and prematurely hailed. However, Cannizzaro noted that only technical information can be transmitted (such as numerical data) from one user to another, and that memes cannot transmit the ideas, beliefs, and greater cultural contexts behind the data. Because of this, the recipient is often left to their own devices to interpret what the meme is actually trying to communicate. This opens the door to misinterpretations, and hence introduces variability into the message.

Grant Kien, a communications professor at California State University, East Bay, examined the societal harm memes cause to our public discourse through a variety of different lenses.[4] In the compilation article composed of several different research papers, he lamented that viral memes are capable of performing lasting harm to our discourse,[4] for a wide variety of reasons which we'll explore below.

Mike Godwin, early internet activist and attorney, observed that meme-based discourse was often unconstructive and rarely ever resulted in the exchange of two parties' cultural ideals.[5] Rather, the conversation (if one could call it that) typically resulted in one side accusing the dissenting voice of Nazism. The Oxford Reference Dictionary, in 2012, made an addition for Godwin's Law. Godwin's

---

4   Kien, Grant. "Media Memes and Prosumerist Ethics." *Cultural Studies ↔ Critical Methodologies*, vol. 13, no. 6, 12 Nov. 2013, pp. 554–561, journals.sagepub.com/doi/full/10.1177/1532708613503785, 10.1177/1532708613503785.
5   Godwin. "Meme, Counter-Meme." *Wired*, 1994, www.wired.com/wired/archive/2.10/ godwin.if_pr.html.

Law states that the longer an online discussion goes on for, the probability of Nazi and Hitler references eventually reaches one. This tactic is underhanded because it is often a way of silencing the voice of dissent. Kien also lamented that such comparisons trivialize and misrepresent the Holocaust. To the common meme user, the Holocaust serves as nothing more than a means to an end of reinforcing their precious echo chamber.

Memes ultimately create an endless loop of unproductive discourse. Despite his pessimism about the ability of meme users to avoid Hitler analogies, Godwin suggested that the internet could seemingly combat the damage done by reckless *memeing* by creating counter-memes. Kien later rejected this claim, noting that by the time this reactionary tactic could be employed, the damage would have already been done. The problem with counter-*memeing* is that it relies on combating a flawed communication medium with yet another flawed communication medium; put more crudely, you cannot fight a *shitpost* fire with another *shitpost*. (As we'll explore later with the Kony2012 phenomenon, counter-*memeing* did not work when the Ugandan government tried it.)

Godwin also warned that anyone on the internet would have the power to affect stock prices. Godwin's chillingly accurate prediction came to fruition in early 2021, when the common meme-hosting subreddit r/WallStreetBets encouraged its user base to artificially manipulate the market price of GameStop. The company GameStop had not been profitable for some time, was fraught with poor employee relations, and was losing out to electronic means of selling games (such as Amazon and console-based online marketplaces where one can download games purchased). Legitimate Wall Street investors realized that GameStop was functionally a weak company and shorted the stock—in layman's terms, betting that the company would continue to decline. However, the self-proclaimed *shitposters* began to post memes about the big-name investors shorting the stock and encouraged their fellow users to create an online ruckus and buy

GameStop shares to inflate the price, solely for contrarianism and to harm the Wall Street investors. However, memes and the resulting meme-based discourse have a short shelf-life, and eventually the price of GameStop dropped from an artificially high price of $320 per share in January 2021 to $40 in February 2021. Institutional investors who shorted GameStop's stock were hurt by this action, and the move sought only to line the pockets of a select few, likely including some undeserving GameStop executives.

As we observed with GameStop, Memes have served as the proverbial baseball hat to signal to other anonymous posters a kind of *Hey, I believe in this cause too!* Logos have served to unite man around a common set of ideals for thousands of years. Even in the modern day, one wearing a T-shirt or a hat with the logo of a favorite sports team can give strangers common ground. It isn't this book's stance that merely taking away the memes will un-radicalize society; the point is that memes have already caused political and social damage and radicalization. Kien discusses that memes and their flawed discourse have caused the many words and common logos (in his term, "signifiers") to become divorced from their original meaning when used in the meme-based discourse. Pepe the Frog is an extreme yet stark example of this.

The variability of memes has led some to believe that memes can channel a diversity of voices. For example, Jose Flecha-Ortiz, a business researcher based in Puerto Rico, found that memes can help explore varying interpretations of a stressful event.[6] However, these varying interpretations quickly converge into a single opinion, thereby resulting in the formation of an echo chamber.

Huntington also explores the concept of discourse, defining it as the "production of knowledge through language." However, it is important to note that memes often only communicate effectively to

---

6   Flecha Ortiz, José A, et al. "Analysis of the Use of Memes as an Exponent of Collective Coping during COVID-19 in Puerto Rico." *Media International Australia,* vol. 178, no. 1, 24 Oct. 2020, pp. 168–181, 10.1177/1329878x20966379.

a preselected group, and their ability to build intellectual and cultural bridges is limited. Memes rely on inside jokes and references that are typically reserved for this particular group, thereby inhibiting their creation of new knowledge. Compared to other more inclusive means of communicating ideas (such as long-form written texts), memes are less effective at building meaningful public discourse. Rather, what is typically observed is that memes are used to reinforce group norms within an already isolated community. While it is possible for those *in the know* to learn something new, such an idea is farfetched. Furthermore, the meme-based discourse is rarely engaging on a wide level, nor it is typically civil.

The divorce of words and symbols from their original meaning in meme-based discourse often occurs in a multitude of ways. As Kien mentions, the meaning of a word or symbol often changes to fit the narrative of the insulated community that is using the meme. A prime example of this is the aforementioned Pepe the Frog, as the *memeing* community assumed the right to turn him into a symbol of hate. Only the selected group can really understand this soft, often malleable language.

Yet this soft, malleable language has its inherent limits. Not only do memes enable the appropriation and alteration of language, but meme users blatantly have the expectation that the meaning of words (and thus language itself) is their plaything, to be altered at their whim, as Kien mentioned. This speaks to the gall and callousness of meme culture as a whole, wherein the heavy users of the medium assume control of language. However, as Kien's article mentions (among other studies), this artificial and constructed set of terminology does not cross the forum boundaries of the meme's posting, therein revealing the flaw of this medium. The medium enables users to blatantly attempt to contort language; however, the medium is unsuccessful at this task on a large scale.

The widespread use of sarcasm within the medium limits its reach as noted by Kien. Kien revealed that challenging ideas (as in, memes that dissent from the in-group's dogma) are met with sarcasm and irony. These two forms of humor, sarcasm and irony, are used by the in-group to dismiss the new idea. Dismissing the idea, and by extension the user who posted the deviant meme, does nothing to add to engaging civil discourse. Instead, it is a tactic used to shut discourse down. Kien mentioned pointedly in the article that "intelligence may be displaced by cleverness."

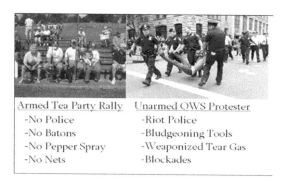

Sarcasm itself is a poor tool to use when not in face-to-face interactions. The use of sarcasm carries the inherent risk of the recipient not understanding the message delivered. That is to say, the other party might take the sarcastic message at face value. Kien's 2013 article mentions his anecdote regarding a meme mentioning the Tea Party protests and Occupy Wall Street protests in 2012, as shown above.

Kien posted this meme to his personal Facebook account, and a friend of his, a Tea Party supporter, earnestly believed that Kien was sympathizing with the Tea Party. Kien professed to have posted this meme in irony. Herein lies the weakness of memes; they rely on sarcasm, which itself is a poor communication medium.

The problem with widespread *memeing* is that once a particular meme catches fire and spreads, it is hard to play damage control. Within his article, Kien uses an example where the original post became an overnight sensation; however, a government fact-check received only 100,000 views. This shows that it can be very difficult to combat lies, slander, misinterpretations, and misrepresentations spread by the common meme. Memes are a reactionary medium; reactions after the fact rarely gain the same amount of attention as the causal event does. Memes are a defensive medium—a medium that, due to the little context they create and the inability to cross cultural lines, is a poor defense.

Memes are often not judged based on truth or reality at all. Kien discusses the troubling phenomena regarding memes; audience members judge memes as a possible means of continuing what he referred to as their "personal spectacle." This is troubling because memes aren't being evaluated on their merits as a communication medium or even if their message holds up to scrutiny. Instead, the common meme is being observed through the lens of *How can I contort this meme to my preexisting worldview?* This is inherently dangerous for informed public discourse if the very channels of mass communication are being hijacked to fit a slanted narrative— doubly so if facts are viewed as irrelevant. Kien continues this point by mentioning that memes are often judged on their aesthetic usefulness. This implies that the lines of text (being a constant, from an aesthetic and font standpoint) are viewed as irrelevant while the image macro is given the proverbial once-over. Memes are such a poor communication medium that their most ardent users do not refer to memes as a tool for communicating ideas.

When one of these memes does become popular enough to warrant a conversation (typically not a long one), it is typically only shared with a preselected "in" crowd. This leads to echo chambers and confirmation bias. There is little, if any, actual exchange of ideas. We observe this phenomenon on Reddit with their tried-and-true downvote mechanic, where voices of dissent can effectively be censored.

Christian Bauckhage, an artificial intelligence researcher at the University of Bonn in Germany, performed a study regarding the survivability of memes.[7] Bauckhage found that memes tended to proliferate within isolated communities instead of crossing boundaries. Memes without significant alteration to their original form do not cross cultural boundaries very well, if at all. Therefore, one cannot reasonably say that memes create an engaging civil discourse or exchange of ideas, given their nature to not stray far from their ideological home.

In a study discussing Israeli meme use during the height of COVID-19, Kertcher and Turin found that memes served mostly as a diversion more than anything else. Immediately before and during the lockdowns, Israel was in the middle of political strife. Their prime minister was facing corruption charges, and the government was facing a shutdown. When COVID forced lockdowns on the nation, the government could provide limited financial assistance, and unemployment shot through the roof. However, most of the Israeli sampled memes didn't address any of these issues. Rather, the memes made attempts at humor regarding trivial things such as toilet paper and social distancing. This indicates that memes do not create engaging public discourse; they merely mock minor problems and divert attention away from actual issues that need to be addressed.

Eva Hernandez-Cuevas, a psychologist from the Kansas City University of Medicine and Biosciences, further quoted Dawkins's work from 2006 regarding memes.[8] Dawkins mentions the existence of both harmful and benign memes. Dawkins further elaborates how benign memes can become harmful due to intentional misuse or misinterpretations. However, as we'll explore later in the book, memes as a medium are rife with ideological slants or messages that are easily taken out of context. Under that circumstance, we can say that most benign memes are susceptible to becoming harmful memes.

---

7   Bauckhage, Christian. *Insights into Internet Memes.* University of Bonn, 2011.
8   Hernandez-Cuevas, Eva Marie. *The Pertinence of Studying Memes in the Social Sciences.* Kansas City University, 2021, www.researchgate.net/publication/352258835.

Dawkins, in 2006 (as cited by Hernandez-Cuevas in 2021), also mentions the presence of beneficial memes. However, given the illusory truth effect that is so common in the medium, one must take a meme that offers supposedly good advice with a grain of salt. However, let's set aside the illusory truth effect for the moment and, for the sake of argument, say that the beneficial memes are proposing useful advice. One of the biggest selling points for the medium is their use of humor, which is not very conducive to their use as public service announcements. Therefore, the very concept of a beneficial meme is limited.

Caroline Drury, in a thesis she wrote while at East Tennessee State University, discusses the use of memes within the emergency medical technician (EMT) community.[9] The author sourced memes about this professional niche from the subreddit r/EMS. The author discusses one particular meme that concerned the long hours the job requires, and then chronicled the resulting Reddit discussion. Rather than a free exchange of ideas, what resulted was an echo chamber. One commentator had replied that they enjoyed their job as a helicopter-based paramedic and loved the benefits and pay. Rather than being congratulated, the happy employee was instead downvoted—their comment was buried by downvoting to the bottom of the page. This is one example of how memes create an echo chamber, by enabling such heavy-handed meme-based censorship. To paraphrase an anti-steroids PSA from the mid-2000s: memes don't create engaging civil discourse; they destroy it.

---

9 Drury, Caroline. *The Function of Internet Memes in Helping EMS Providers Cope with Stress and Burnout.* 2019.

ARE MEMES EDUCATIONAL TOOLS?

MEMES OFTEN SERVE as the PowerPoint slides of the internet. PowerPoint slides have been an inferior means of communication and education. A widely read *New York Times* article written by Elizabeth Bumiller in 2010 highlights the frustration of PowerPoint reliance.[1] Instead of covering a boardroom meeting or an education conference, the articles covered the use of PowerPoint within the United States war effort in Afghanistan. General James Mattis of the United States Marine Corp (USMC) voiced his disdain for the tool, claiming "PowerPoint makes us stupid." General Mattis wasn't alone in his sentiment. Also featured in the article is Brigadier General H. R. McMaster, who said of PowerPoint, "It's dangerous because it can create the illusion of understanding and the illusion of control." He added "some problems in the world are not bullet-

---

1   Bumiller, Elisabeth. "We Have Met the Enemy and He Is PowerPoint." *The New York Times*, 26 Apr. 2010, www.nytimes.com/2010/04/27/world/27powerpoint.html

izable." McMaster's disdain for the tool lead him to ban the use of PowerPoint altogether when he lead the campaign to successfully retake the Iraqi city of Tal Afar in 2005. The bullet-izing of nuanced, multifaceted issues was causing people to believe they understood an issue when there was proof to the contrary. Stripping away vital context to fit neatly into a limited space carves out context that is often vitally necessary.

The use of PowerPoint slides "stifles discussion, critical thinking and thoughtful decision-making," as Bumiller mentions in the article. At their worst, PowerPoint slides are an ineffective communication medium. However, all hope is not lost for the Microsoft application; a study was performed at the University of Minnesota Veterinary School that demonstrated how to make a better PowerPoint presentation, one that improved the retention of viewers. Veterinary students were split into two groups, one with traditional slides (that is to say, laden with bullet points that strip away many of the content) and one with what the study calls assertion-evidence slides (large headline, one sentence, and a visual aid). Students shown the assertion-evidence model of PowerPoint slides scored higher on test scores by a statistically significant margin; the test was designed to demonstrate long-term retention of the material being taught.[2]

One important thing to note here is that PowerPoint slides are *presented.* This example occurred in a veterinary school, where a qualified professor with decades of experience carefully delivered the material, delivering context and shaping the nuance of the material to lodge the knowledge deep inside the brains of high achievers. Even in this best-case scenario, the communication medium of PowerPoint slides cannot survive as a stand-alone communication medium and are inherently reliant upon a knowledgeable expert to cautiously

---

2   Root Kustritz, Margaret V. "Effect of Differing PowerPoint Slide Design on Multiple-Choice Test Scores for Assessment of Knowledge and Retention in a Theriogenology Course." *Journal of Veterinary Medical Education*, vol. 41, no. 3, Sept. 2014, pp. 311–317, 10.3138/jvme.0114-004r.

create the background necessary. As the evidence suggests, generic formatting is not an effective way of getting a point across.

Memes are *posted,* not presented like PowerPoint slides. Memes seldom have the benefit of a benevolent instructor to guide an audience. In most cases, memes are consumed in a vacuum and stripped down to the bare minimum (or even less) to make a coherent point (assuming on is made at all). Memes, therefore, are entirely dependent on the audience being being worldly enough to grasp their foundational concepts. As discussed, PowerPoint slides (even at their best) are still stripped of vital context and are reliant upon an expert to fill in the gaps. Memes are inferior to PowerPoint slides in that they aren't usually paired with an external expert.

"Memes aren't facts," as comedic journalist John Oliver informed the masses on his hit show about vaccine skepticism.[3] In each episode of Oliver's show *Last Week Tonight,* Oliver closely examines a pressing social or political issue (albeit with a left-of-center bias, which Oliver has admitted before). Oliver's show is hosted on the HBO platform and is flush with financial and investigative resources. With so many means of seeking the truth at his disposal, Oliver wisely decided to cast memes aside when seeking legitimate sources of truth. Independent of whatever stance you take on his social or political views, John Oliver is a man of character for refusing to use memes.

Memes, as explored by Kien, enable their consumers to feel comfortable seeking answers to profound questions. Such comfort yields a false sense of confidence because the creators and posters of these memes likely do not have specific expertise that viewers can rely on. Nor, as opposed to John Oliver, do they seek out expert insights. Even if a meme portrays a truthful answer, the forced brevity of the broken communication medium prevents the creator/poster from

3   Vaccines: Last Week Tonight with John Oliver (HBO)." *Www.youtube.com,* John Oliver, June 2017, www.youtube.com/watch?v=7VG_s2PCH_c&t=988s. Accessed Aug. 2021.

adding any context to explain why their answer is correct. Without this context, it can be hard for those seeking answers to parse truth from lies.

NationalReportCard.gov is a website that houses a graph displaying the standardized testing scores of our nation's youth (same ages/ranges). The website features an interactive map where a user can select two comparison years and determine the score variance between the two. As the graph below will demonstrate, eighth grade students in most of our nation have either stagnated or gotten *worse* at reading in 2019 compared to 2002. The year 2002 serves as a good proxy for a pre-meme society as it is before the advent of many popular *memeing* sites. (Disclosure, this data set is also lumping in the District of Colombia and military children who are educated through the Department of Defense Education Activity, bringing the total up to fifty-two jurisdictions rather than fifty, though that does not alter the trend described.) A snapshot of their findings is listedon the next page.

Between

2002 and 2019

7 states/jurisdictions ←	23 states/jurisdictions ◆	13 states/jurisdictions →	9 states/jurisdictions ⊘
had a score increase	had no significant change in score	had a score decrease	had no data or not applicable

NOTE: DS = Department of Defense Education Activity (DoDEA), a federally operated nonpublic school system responsible for educating children of military families. See more about DoDEA.

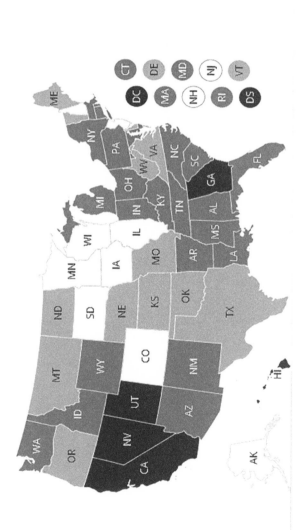

Additionally, the average SAT critical reading scores of our nation's college-bound youths has similarly decreased from the year 1996 to 2016.[4] The average score in our pre-memetic society of 1996 was 505, while in 2016, the last year of the SAT format of that time, had slid all the way down to 494.

But why do you keep bringing up our nation's reading scores? And what relevance does this have? a misguided meme defender may ask. The point is simple; while correlation doesn't necessarily prove causation, we would be utterly naive to believe that the nations slide in reading scores (a useful proxy for critical reasoning and attention span) has occurred completely independent of the rise of memes. Surely, the fall of our nation's reading scores is a multifaceted issue; however, only the most naive among us would deny that meme use, and the damage they cause to our communication and reasoning skills, wasn't a factor in the decline. When Reddit accounts have become more ubiquitous among American youth than library cards, there is something amiss.

The evidence doesn't just exist in the United States. A Japanese team surveyed schoolchildren regarding their frequency of internet use and then used a magnetic resonance image (MRI) machine to scan their brains.[5] The study revealed a strong negative correlation between internet use and white matter in the brain. That is to say, the more often they use the internet, the less white matter their brains developed. A decline in white matter is concerning because white matter is where the most synaptic firing occurs, i.e., where messages are sent from one neuron to another. White matter derives its name from the heavy presence of myelin, a sheath that wraps around the axon of a neuron that acts as an electrical insulation. Without this precious insulation, electrical signals cannot propagate down from one neuron to another.

4   Aldric, Anna. "Average SAT Scores over Time: 1972 - 2019." *Prepscholar.com,* 2021, blog.prepscholar.com/average-sat-scores-over-time.
5   Takeuchi, Hikaru, et al. "Impact of Frequency of Internet Use on Development of Brain Structures and Verbal Intelligence: Longitudinal Analyses." *Human Brain Mapping,* vol. 39, no. 11, 28 June 2018, pp. 4471–4479, 10.1002/hbm.24286.

Furthermore, the scientists found that the most underdeveloped areas of the student's brains were the centers that process emotional response and the reward feedback loop. Other impacted areas of the brain involved centers that processed language development, memory, and attention. A particularly troubling finding is the associated drop in verbal IQ with frequent internet use.

While the study did not speculate as to whether or not memes were consumed, other data suggests that they were. Data mined by website traffic and search results by ahrefs.com, a popular tech blog, reveals that the common meme-peddling websites are very well represented both domestically and internationally, as of January 2021. Facebook, Reddit, Instagram, and Twitter all crack the top twenty websites used in both categories. So, it isn't a complete stretch to say that the youths participating in the study were visiting these websites (given that many of these sites allow users below the age of eighteen to access their sites).

Memes aren't just impacting the kindergarten through twelfth-grade classroom in America; they are also being used in higher education. The University of Florida's pharmacy course used memes in one of their courses for third year students taught by Dr. Joshua Brown.[6] This study exemplifies some of the trouble with memes as an educational medium. The professor admitted to finding it difficult to use memes as an educational tool due to trying to fit the course material into the construct of a meme. Memes are notoriously bad at building context, which is a vital part of the learning process. This is especially true within a field like pharmacy where *patients' lives are on the line*; one would hope that the pharmacist has a solid conceptual grasp of the material. Memes are not a medium that easily allow for this, even in the hands of a caring instructor.

Brown placed the students into groups and had them create

6   Brown, Joshua D. "What Do You Meme, Professor? An Experiment Using 'Memes' in Pharmacy Education." *Pharmacy*, vol. 8, no. 4, 29 Oct. 2020, p. 202, 10.3390/pharmacy8040202.

memes about the course material. Brown observed that material that was less important to the course was overrepresented in the memes submitted. One example of this was that students tended to create memes regarding the professor's quotes or often tangential subjects that had very little to do with the course material. Memes divert learners away from the topic at hand, often to wind up in the solace of something much more trivial. Brown had his students continue the meme assignment several times over the semester and thus observed that students repeatedly misrepresented core concepts over several group assignments. When pharmacy students (who themselves are high achievers with startlingly high grade point averages in undergraduate work) begin to meme, even they are prone to getting off track. Memes are not an ideal medium for serious academic work.

Brown acknowledged that, for the meme assignment to be considered successful, more context and tighter control was needed. A greater set of restrictions was necessary to prevent the wandering of students into off-topic areas. This is seldom a problem with long-form written essays submitted by graduate students, as the question is clear and the medium is conducive to displaying true understanding of the material. Memes are brief communication mediums that often belie themselves to needing information shoehorned into them, as they are fundamentally unable to explore long or difficult concepts.

Brown admitted in his own paper that using memes to teach course material would be a tough sell to his older colleagues. Generation X professors or older may have a difficult time seeing the educational value in memes and (as Brown feared) would avoid the medium in the classroom. Given the evidence presented, it would certainly be difficult for Brown's colleagues to be judged harshly for that stance. Memes are simply not a medium that avails itself to teaching. Memes lack context and are used to police old behavioral standards rather than instruct new ones. Both of these are serious pitfalls in an activity in which knowledge transfer is involved.

Brown also lamented the inherent shortcoming of the common meme as a teaching tool—the need to constantly stay on top of meme trends. Say, for the sake of argument, Professor X introduces meme-based learning in his hypothetical classroom. Brown states that the onus is on Professor X to stay current on the meme trends in order to effectively communicate with students and to build a rapport. To do this, it stands to reason that Professor X would need to dedicate precious time to staying up on the trends of the *memeing* community. This would naturally take time away from more traditional responsibilities such as lesson planning, grading, researching relevant advances in his field, etc. Should Professor X neglect this newfound duty, he risks alienating his students with the use of memes, much like (to quote Brown) "references to 'out of date' or 'old school' musicians or television shows often do." Using memes, therefore, would counteract the rapport building that Professor X originally set out to accomplish in the first place.

Brown admits further limitations in the use of memes as a teaching method. Introducing memes into the course often blurred the lines between formal and informal environments, which are essential to the boundaries of professionalism. As we'll observe in the study by Rabea Hecker (a thesis written while a communications student at the University of Twente in the Netherlands), memes make the viewer less likely to take the subject matter seriously.[7] This isn't a major concern for students in Art History 101; however, these stakes are considerably higher with students in the field of medicine.

Memes have not only been utilized in the graduate spaces, but they have seeped into the undergraduate levels of higher education as well. Rishabh Reddy, an engineering professor in India, introduced memes into several undergraduate engineering courses

---

7   Hecker, Rabea. "Are You Serious It Is Just a Joke? The Influence of Internet Memes on the Perception and Interpretation of Online Communication in Social Media." 2020.

at a university in India.[8] In the introduction of the paper, Reddy concedes that memes are only an effective communication medium (and therefore, an effective teaching medium) if the full meaning of the meme is understood. This involves the recipient not only realizing how the imagery connects to the short, quippy phrases but also being able to evaluate the semantics of the phrases themselves, i.e., for both legitimate content and attempted humor. Without this full understanding, memes in this context become a poor tool for their intended purpose.

Reddy further concedes that memes need to be funny in order to have the desired educational effect. Stated pointedly, if a meme is factually correct but fails to generate interest, then it fails at its fundamental task in the classroom. The memes created for educational purposes, therefore, are forced to omit content in order to generate cheap laughs. Reddy acknowledges that a balance must be struck between content and humor when creating educational memes. We must ask the rhetorical question, *Would you be comfortable with a teacher skimping on vital content if they were teaching your kids?*

In the methodology section of the research paper, Reddy further concedes that memes do not directly impact student performance. That is a crippling concession to make when the entire point of the paper is to evaluate the role of memes in the classroom. Teaching tools are intended to be implemented for this very purpose, and if they are not fulfilling this purpose, then one must seriously ask why memes are being used in the classroom in the first place.

Reddy then polled students in the meme-based classroom for the opinions after the course ended. Many of the students polled viewed the usage of memes favorably, with 63.5 percent of students feeling that the use of memes helped them understand the course material better. However, there is a methodological flaw in the study of memes

---

8   Reddy, Rishabh, et al. "Joy of Learning through Internet Memes." *International Journal of Engineering Pedagogy (IJEP)*, vol. 10, no. 5, 15 Oct. 2020, p. 116, 10.3991/ijep.v10i5.15211.

based in the education field; student opinions are *not the same* as student results. It is a very important distinction to make, and many studies revolving around the use of memes in the classroom tend to consider student opinions to be valid data.

Reddy also discussed another Achilles' heel of memes—the fact that they divert student attention away from the material at hand. Students in the undergraduate courses took the course less seriously through the extensive use of the meme. How can any serious instructor expect students to learn the material if students do not take the class seriously? This leads to the same blurred lines of professionalism and formality that Brown discussed regarding his students at the University of Florida. Given that this phenomenon has been shown to occur in two different cultural environments (United States and India) with high achieving students (pharmacy students and engineers), this bolsters the case against memes.

Reddy concludes the paper with the admission that memes are not an innovative tool used for teaching. The author further concedes that memes are not a powerful tool for teaching engineering courses. The paper makes the concession that memes aren't even useful for making the material easier to understand. Despite the fact that the paper is titled "Joy of Learning Through Internet Memes," the paper spends a lot of time distancing itself from claims regarding the usefulness of memes in learning.

A final educational study of memes reinforces a lot of Brown's and Reddy's findings, despite claiming the usefulness of memes in the classroom. Giulia Bini and Ornella Robutti, two Italian mathematical researchers, studied the use of memes with a high school math class in Italy.[9] The study runs into many of the same methodological shortcomings and reveals (intentionally or otherwise) many of the weaknesses of memes in the educational arena.

---

9   Bini, Giulia, and Ornella Robutti. "Thinking inside the Post: Investigating the Didactical Use of Mathematical Internet Memes." *International Group for the Psychology of Mathematics Education*, 2019, hdl.handle.net/2318/1698304.

Bini and Robutti begin their paper by giving some background into what a meme fundamentally is, exploring the three concepts that give a meme meaning, the first two being the structure and the sociability of a meme. The authors of the paper conceded that many instructors (inherently from a different generation than students) would have a difficult time using memes effectively in the classroom due to not fully comprehending the fundamental concept of a meme. Given the evidence in that paper, the meme-based learning is only effective when all three concepts are present. This phenomenon, an inherent barrier between teacher and student forms, limits the learning process. Memes are inherently limited in this use.

Bini and Robutti mentioned that memes are used to hit the emotions of students and to bias attention; however, an instructor must be careful when attempting to bias the attention of the students using memes as a medium. Memes, as we saw in the University of Florida study and the Reddy study, can cause students to not take the class seriously (an inherent bias of attention). Further, as we covered in the previous paragraph, many instructors are limited in how effectively they can use memes in the classroom. This is an inherently dangerous combination.

Bini and Robutti noted that memes find their reason for being in reactions, and that memes are rooted deeply in emotions. So, memes as a communication and educational medium are so deeply flawed, it's rarely about the message at all; rather, it's the meme itself that is driving these responses from these students (and recipients as a whole). The message, the actual *educational* part of the classroom, is not relevant, given this premise. Stated earlier in this chapter, memes that are correct but not humorous will not be a functional meme. We must ask ourselves as a society—are we okay with memes being used to emotionally manipulate the youth, especially when the message is light on content? There are more effective means of delivering and reviewing course material to students.

Students in the Bini and Robutti study were asked to create

memes relevant to the course material. One pair of students avoided using a well-known meme format for the exercise. When asked why, the students admitted that they wanted to differentiate themselves. The researchers had added that templates that had a lower likelihood of gaining *likes* or other external means of validation were less likely to be used. This shines another spotlight upon the weakness of memes; this is a medium where only the *dank* survive. Shareability and ability to generate cheap gimmick laughs were absolutely paramount while learning, and personal growth fell to the wayside, which is the opposite intended effect of a classroom.

Memes are only effective if the full meaning is understood. Therefore, memes are best suited for two practicing professionals sharing an inside joke, rather than as a vehicle for introducing new content to unknowing audience members. Memes, at best, are a means of reinforcing material already covered rather than introducing new material.

Memes cannot stand alone to deliver novel content to an audience. Much like we saw in the University of Minnesota's Veterinary program, memes are a weak medium that need to be propped up artificially by the guiding hand of a caring expert. Some websites, such as Khan Academy, do not have this problem and are clearly a superior choice to education than internet memes.

In summary, memes have been demonstrated to cause students to not engage with the material in a serious manner, have had researchers make several ominous concessions regarding their effectiveness, are tested by studies rife with methodological flaws, and cannot be used to teach new material (as a prior level of understanding is assumed with a meme). Therefore, one can confidently state that memes do not belong in the classroom.

MEMES AREN'T ADDICTIVE... ARE THEY?

A PEW RESEARCH POLL from 2018 (and then repeated in February 2021) shows that the common meme-hosting websites are well represented in everyday internet use.[1] Nearly one in five adults use Reddit (18 percent) and TikTok (21 percent), nearly a third use Pinterest (31 percent), while two out of five partake in Instagram (40 percent), and the undisputed king of social media, Facebook, clocks in at nearly seven in ten American adults who use the website. Meanwhile, a separate Pew study reveals that nearly half of Facebook (49 percent) and Twitter (46 percent) and nearly three out of five Instagram users (59 percent) use those meme-propagating platforms

1   Pew Research Center. "Social Media Fact Sheet." *Pew Research Center: Internet, Science & Tech*, Pew Research Center, 7 Apr. 2021, www.pewresearch.org/internet/fact-sheet/social-media/.

"several times per day."[2] These websites have been designed to addict us—and they succeeded.

Memes are also detrimental to the social dynamics of America's neighborhoods. Seemingly gone are the days when neighbors engage one another over fences and when children and teenagers socialize with their peers outside after school. Internet use has skyrocketed to the point where addiction to the internet, while not officially recognized by the leading medical catalogs, has spawned an industry of counselors delivering therapy to those who cannot seem to separate themselves from the internet. It is no secret that many of today's top websites have designed themselves to engage users by delivering a dopamine hit to users on a regular basis.

Common meme-hosting websites such as Facebook, Reddit, and 4chan are prevalent in the group of sites known for addicting their users to their platforms. Hours upon hours constantly fixated upon the latest radicalized political meme, Minion meme, or popular circulated meme takes away from genuine human connection, such as familiarizing oneself with their neighbors and peers. Shuttered inside and focusing with a pseudo-community online, adolescents and teenagers lose out on important developmental interactions that come with playing sports or joining other activities that are commonly offered as after-school clubs, such as board games, playing musical instruments, creating art, and more. Alas, the social costs of internet addiction are steep.

Minors, however, are not the only ones at risk. Millennials, now in their adulthood, often engage in *memeing*. Entire social outings spent with eyes glued to their phones or mindlessly scrolling Imgur, a website that hosts custom-made photos and memes, is not conducive to building social skills and engaging with the wider world. *Memeing* often happens on the job, resulting in decreased productivity. The

2 Gramlich, John. "10 Facts about Americans and Facebook." *Pew Research Center*, 1 June 2021, www.pewresearch.org/fact-tank/2021/06/01/facts-about-americans-and-facebook/.

average American employee, despite spending more hours at work than their parents' generation did, often spends less time performing actual work. The resulting decrease in employee productivity can, and does, have real impacts on the wider economy. When employees are less productive overall, companies are forced to lean more on the more productive employees to pick up the slack (resulting in burnout and therefore higher turnover). The resulting higher turnover is a blow to companies, as it is often expensive and time-consuming to replace a productive employee. Companies can also be forced to raise prices, thereby passing the burden to consumers.

As a defensive measure, many employers often block common meme-hosting websites such as Facebook, Instagram, Reddit, and 4Chan on company-owned computers, much to the dismay of employees; however, one cannot necessarily blame the companies for wanting to do this. Employees are paid to work, not to scroll through memes on the job. Visiting these websites during company time is often disrespectful to the company and flies in the face of the fiduciary responsibilities of the investors of the company. In short, *memeing* may not directly fraud investors or customers, but the lost productivity stemming from their use definitely cannot (and should not) be ignored.

Before we can explore how common meme-hosting websites design their platforms to cause addiction, we must first give a brief overview of the science behind addiction. Two brain cells (neurons) have a small gap between them called a synapse. In order for one neuron to "talk" to another, that neuron must send a molecule known as a neurotransmitter across the synapse to the receiving neuron. The sending neuron must transmit a minimum threshold of transmitters to the receiving neuron. Once this minimum threshold is met (known as an "action potential" and measured by electric charge), the receiving neuron then transmits to the next neuron. All of this happens within fractions of a second.

Dopamine is a neurotransmitter that serves a key role in the reward and motivation feedback loops of the brain. Several addictive drugs either increase the amounts of dopamine released or block the reuptake of dopamine (therefore making dopamine persist longer). As more dopamine is chronically released, the brain adapts by creating more dopamine receptors. Therefore, more dopamine is needed to feel the same phenomenon as the user previously felt. This, in a nutshell, is the science of addiction.

Common meme-hosting websites understand this feedback loop and have introduced validation features, such as *likes* and *hearts*, to reward the poster and release dopamine. They further added a comments section with a notification service so, when another notification is received, it results in another dopamine hit. The posters brain adapts, and now requires more validation from their misguided online community. The poster has become astutely aware to what constitutes a validating post and thus becomes heavily vested in their subculture. Now, overtly ruled by their feedback loops, the meme poster is addicted to the website, and thus, provides a captive audience and a source of data generation for the Silicon Valley elite.

In a study we'll examine in depth later, Karatozgianni, Miazhevich, and Dennisova talk about the communities that form with the use of meme-based discourse. They argue that the meme posters do not need to belong to a large forum or community. Alone and isolated, meme users continue to radically spew their memes into public discourse, hoping to pull others with them. This clearly is not a healthy way to communicate with the outside world. The fact that meme users do not need to belong to a larger group supports the argument that memes are addictive. If memes did not cause addiction, then their most ardent users would not need to self-isolate.

Cannizzaro wrote a scholarly article regarding the use of internet memes through the lens of semiotics, which is the study of signs and signals. In the very abstract of the paper, Cannizzaro states that "memes can and should be conceived, then, as habit-inducing."

The academic community willingly acknowledges that memes as a medium are addictive. Given what we have explored earlier in the education chapter regarding the generational gap within *memeing*, it can be deduced that memes are addicting the youth of this nation (and the youths of many other nations) at an alarming rate.

Cannizzaro details the brief history of the "Grumpy Cat" meme, describing how the meme was then used in a British TV commercial for a bank (a for-profit institution). Marketers are constantly trying to come up with advertisements that *stick*, which is a quality of an addictive medium. Since memes are used by marketers for this purpose, it isn't a radical leap of faith to say that memes have a habit-forming quality to them.

Umair Akram, a psychologist at Sheffield Hallam University in the UK, used internet memes in a study of depression.[3] The study consisted of volunteers who were sectioned off into two groups, a non-depressed control group and a group of depressive patients. Each participant observed thirty-two internet memes, sixteen depressive memes and sixteen control memes (memes that had nothing to do with depression). Retina-tracking technology was used to track movement in the participant's eyes, the initial gaze, and the total retention time. Basically, the software could detect who focused on what meme, and for how long. From that, the study found that depressive patients fixated longer on the depressive memes than the control memes. The results also showed that depressive patients paid far less attention to control memes than they did the depressive ones. Given this, there is an inherent danger with internet memes as a communication medium. Since memes are often used as thinly veiled jokes, the danger is that someone undergoing emotional trauma will seek out this medium, perhaps unaware that the medium wasn't trying to be serious.

---

3   Akram, Umair, et al. "Eye Tracking and Attentional Bias for Depressive Internet Memes in Depression." *Experimental Brain Research*, vol. 239, no. 2, 17 Dec. 2020, pp. 575–581, 10.1007/s00221-020-06001-8.

Thankfully, as we'll explore in the activism chapter, memes do not encourage people to act upon their content in the real world. While the ineffectiveness of the communication medium serves as a de facto safeguard in that aspect, it still isn't ethical to pull depressive patients down a spiral of negative emotions. This is a failure of memes, because without the meme's ability to create context, the viewer is often left to interpret the meaning. A longer-form communication medium would certainly create more context, as the author would be able to clearly delineate between satire and earnest.

Amanda du Perez and Elaine Lombard explored how memes spread, in part by appealing to positive or negative emotional content. The very means of the communication medium's way of surviving is to manipulate the emotions of the reader. Memes, as a medium, in this aspect, are not an honest or straightforward means of communication. If memes were a robust communication medium, they would not have to rely on raw manipulation to begin with. Memes, for this reason, simply cannot be trusted.

Amanda du Perez and Elaine Lombard further discussed how the memes one chooses to represent themselves online typically don't require a lot of scrutiny. A meme user typically does not consciously know (or at least is unlikely to consider) the consequences of what they are posting. This is indicative of how habit-forming the communication usually becomes. Memes as a communication medium do not usually encourage the user to stop and think about their postings. The addiction to memes and likes and validation push the user to post without an understanding of the consequences; those are usually an afterthought, if they're a thought at all.

THERE'S NOTHING WRONG WITH A FUNNY MEME...
RIGHT?

A COMMON ELEMENT found in the flawed communication medium is humor. Humor can be a great means to convey an idea, if used selectively. But the meme world relies on humor to such a degree that few other emotions (except for anger) are ever truly used in meme-based dialog. Why do memes feel the need to be funny? Like spices, humor can provide a necessary bit of flavor to the grander dish. We must keep in mind that humor, again like spices, can be overused, and that is very much the case with memes.

As we've explored in the education chapter, humor is often employed as a medium for increasing the recall of information. This has led to internet memes using humor (overwhelmingly so) in hopes of conveying a message. And as explored in the education chapter, students did recall some facts due to the use of humor (albeit the facts that were recalled were dubious at best in terms of relevancy). Repetition breeds believability regardless of truth, making a humorous (though untrue) meme's message stick. Once the message sticks with the user, it is more likely to be spread. This can further *parasitize* the minds of others, to borrow a concept from Richard Dawkins, the original theorist of memes.

With that said, it isn't difficult to see that those with an agenda would value the common meme. The meme, in its common form, lends itself well to spreading misinformation. All the creator has to do is package their devious message in a humorous meme, and their message can spread exponentially (or live on as an often-recycled meme). It isn't the stance of this book that all humorous memes do this, but it *is* this book's stance that all funny memes are susceptible to this pattern. The meme pictured below illustrates my point.

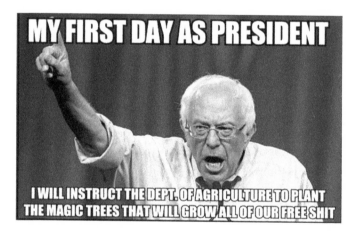

This meme depicts longtime Vermont senator Bernie Sanders passionately giving a speech.[1] The humor comes in the bottom text, alluding to the "magic trees" that will produce the policies that Bernie has spent most of his career advocating for. This humor has been marketed toward an audience who is unlikely to vote for Bernie Sanders. The punchline of "magic trees" uses sarcastic humor to underscore the opinion that Bernie Sanders' policies are unobtainable. Now, had this user written a long-form essay/blog

---

1 smooveb. "SAVAGE Political Memes 3." *Www.ebaumsworld.com*, www. ebaumsworld.com/pictures/savage-political-memes-3/85159074/. Accessed 2021.

posting about the difficulties of obtaining Bernie's ideal society (complete with valid scientific and economic references), it wouldn't have the humorous punch, nor the resulting spreadability. Doing so would have created a solid foundation for their argument. Humor was this meme creators' ideal tool to proliferate a message about their least favorite candidate. Basically, the use of the humorous meme caused the poster (again, note how the term *creator* was not used) to omit crucial context and reduce their likelihood of convincing others of their anti-Bernie views. In order to avoid the accusation of playing partisan politics, we'll examine another meme that uses humor to spread an agenda from an opposing viewpoint.

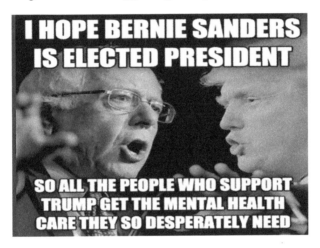

This meme depicts Sanders and Donald Trump juxtaposed.[2] The humor comes once again from the bottom of the image. The punchline implies that Donald Trump supporters are mentally ill and thus would benefit from Sanders' universal health care proposal. Once again, a well-researched and defended book or published article wouldn't spread as quickly as the common meme and, of course, is far more difficult to accomplish, as there is an inverse

---

2 Giovanni. "The Best 2016 Political Memes." *Urban Myths*, 5 Mar. 2016, www.urbanmyths.com/urban-myths/politics/the-best-2016-political-memes/. Accessed 2021.

relationship between ease of accomplishment and skill required for said accomplishment. However, a longer form and well-referenced medium would have created a much stronger basis for their argument. Furthermore, the meme fails to state why the creator feels Trump supporters are mentally ill, as memes are a poor communication medium of ideas. This medium encourages their users to sacrifice substance for spreadability, and that is the overwhelming problem with memes.

Yus also explored the role of humor in the journey of the common meme. As supported in another study by Blyth Crawford, Florence Keen, and Guillermo Suarez-Tangil (three anthropologists from Kings College in London), humor is used as a means of cloaking the dark and hateful content of a meme, when memes are in the hands of a radical.[3] Yus discussed that humor also serves as a way of spreading from one user to another. Yus further discusses that if the meme can make the user feel good in some way, the meme is much more likely to spread. These two findings juxtaposed to one another should make one wary next time they see an internet meme attempting humor. Worse yet, the humor used in the common meme is typically that of a simple nature. These types of jokes often rely on simple juxtaposition, compare/contrast jokes, or multiple definitions or context of a word, as Yus demonstrates. Once again, the brevity of the medium forces compromises to be made.

Another aspect of humor that isn't commonly discussed is the repetitiveness of the joke itself. The repetitive use of the same jokes become less and less funny with each retelling, as the novelty wears off. Memes are rife with lowbrow attempts at humor that fall prey to the same tired, old discussion points. We'll examine an example of this below.

---

3 Crawford, Blyth, et al. "Memes, Radicalisation, and the Promotion of Violence on Chan Sites." *The International AAAI Conference on Web and Social Media (ICWSM)*, 7 June 2021, kclpure.kcl.ac.uk/portal/en/publications/memes-radicalisation-and-the-promotion-of-violence-on-chan-sites(ec3ef161-783e-4403-a09d-4858f42647df).html. Accessed 24 Nov. 2021

This meme is called the High Expectation Asian Father.[4] To anyone who has used the internet since 2008, this meme is certainly nothing new. The premise preys on the worn-out stereotype of Asian-American parents being strict regarding the grades of their children. This cultural stereotype has been paraded around in the United States for several decades. This joke offers nothing new or informative, and mostly serves as cheap humor at the expense of an entire category of people. The joke is also tacky in its use of intentionally poor grammar, seemingly to mock the Asian father portrayed as having poor English. The use of memes encourages this lowbrow form of entertainment, which, in this case, relies on blatant racism.

Philosophically speaking, why have we accepted cheap entertainment like this? At what point did we say that discourse like this was humorous? Nearly every social media website prides themselves as a platform where ideas can be spread and exchanged, so why are *these* the ideas that we have chosen to spread?

One common way for memes to spread is to use humor as a guide for subversiveness. Memes serve as a type of visual rhetoric, a quick way to signal to others in one's in-group. Typically, this in-group holds ideals that run counter to those of polite society, whether it

---

4 "High Expectations Asian Father." *Imgflip*, 2011, imgflip.com/i/m2. Accessed 2021.

be a social or political cause. Many times, the humor used in these memes can be simple juxtaposition or hyperbole; longer forms of comedy, such as tactics deployed in stand-up, take too long to build up to, and the skill barrier is too high.

The subversiveness of the common meme is explored in a study on extremist memes on 4chan and the now defunct 8chan (now known as 8kun). The Crawfrod, Keen, and Suarez-Tangil study observed that part of the culture of these meme-hosting sites was to post "shock memes"—memes so grotesque and violent that they would shock a member of civil society. One such example of this cited in the Crawford article depicts the aforementioned Pepe murdering a Jewish merchant, done so with heavy alt-right undertones. These memes were often endorsed by their ideological cohorts on these boards; their celebration derives from their mainstream cultural subversion.

The urge to use humor as subversion represents a flight not only from decency, but from reality itself. Drury examined the use of memes as a communication medium (more specifically as a coping mechanism) among emergency medical technicians. The author discussed a meme that was posted to the subreddit r/EMS describing transporting mental health patients. One commenter, a former psychiatric patient, had felt the need to respond to the meme, which they felt had gone too far. They recounted their own experience of not wanting to appear threatening to the EMTs, afraid to move after being told they could adjust their position for comfort. The original poster (allegedly an EMT) admitted to exaggerating the situation for the sake of the meme. The need to include humor was so powerful that the poster was willing to sacrifice integrity for the sake of their meme. Not only is this dishonest, but it leads to the stigmatization of mental health patients, from a profession that is supposed to have their best interests at heart. This goes to show that memes are not a dependable source of communication and cannot be trusted to accurately represent a story. The old adage *art imitates life* may hold true, but we need to add the suffix *but memes do not.*

THE HYPOCRISY OF MEMES

TAKE A LOOK at the online Reddit community of r/DankMemes and one will find (along the right side of screen) a lengthy list of rules that the moderators have compiled to govern their community's behavior. Some common regulations include a limit to how often one can post to this subreddit and a plea to not spam the subreddit. Take another stroll over to r/Memes and one will find a similar list of rules about not spamming their precious space. Herein lies the hypocrisy of the medium; memes are created and proliferated throughout the internet, to infest the feeds of innocent bystanders everywhere. Yet the posters *hate* when it's done to them. This inherent contradiction of what these communities preach versus what they practice is jarring.

The r/DankMemes subreddit has several different flairs (labels) for the header of each meme posted. Flairs serve to properly categorize the meme. One such flair is *low-effort memes*. This is redundant due to the aforementioned lack of effort it takes to create a meme to begin with. Given that and the prevalence of lowbrow humor used in the medium, what *is* a high-effort meme? Without knowing the criteria these gatekeepers use, nearly any meme can be classified as this category. It takes not a great deal of research nor knowledge to create a meme. Keeping in mind the ease of meme creation and subsequent plagiarism culture of common meme-hosting websites, nearly every meme is a low-effort meme.

A generation or two ago, an organization that wanted to mislead the public had quite a few hurdles to clear before doing so. A message needed to be created, typically in a long-form medium such as a newspaper article. The article had to make it past the editor of a newspaper, who typically followed some sort of journalistic integrity standards. However, for the sake of argument, let's assume that the editor allowed a devious article to be printed. Outside of overarching metrics, such as sales of that particular newspaper on that particular day, there was no discernible way of connecting that article to the public reception (and the placement of the article in the newspaper was also a factor). They typically needed to spend quite the considerable sum of money to spread a devious message to the masses.

Contrast that with today, when an anonymous poster can copy and paste a meme he found on Reddit and post it onto 4chan with little to no guardrails to prevent it from circulating. After quickly going viral on both Reddit and 4chan, someone with malicious intentions can post that same meme to Facebook, where it can further spread. Thanks to the illusory truth effect, the sheer repetition of whatever their message was can now be easily digestible by the masses, all without having to go through a single editor or having access to a large marketing budget.

Memes have also been used by conventional companies for their marketing purposes, though results have been mixed. Reddit has historically been one tough nut to crack. Like most meme-hosting sites, Reddit has a heavily insulated user base. Redditt's downvote feature serves as a de facto censor to any news or positions that are deemed inauthentic or inconsistent with the subreddits norms. Herein lies the hypocrisy; memes are a marketing tool, yet many meme-users hate when the medium is used on them for such a purpose.

If an external marketer can penetrate the Reddit hivemind, there can be bountiful treasure awaiting, as Reddittors are known to be large spenders (relative to their demographic) on technology and easily influenced.[1] Moreover, memes are perceived to be more authentic than other communication mediums, partly because of their anonymity.[2] This, of course, opens the door to more outsider manipulation, which is exactly what the marketers want. Once the Redditt hivemind has been penetrated, memes are an effective way to emotionally manipulate the largely White, male, millennial, early tech-adopter audience to separate themselves from their wallets.

A common term used in academic literature revolving around memes is *prosumer*. This word, which is a portmanteau of *producer* and *consumer*, warrants further scrutiny, though such scrutiny is largely absent from academic studies on the subject of memes. The second half of the word, *sumer* coming from the word *consumer*, seems appropriate enough, given the high *consumption* of memes in our public discourse. However, the beginning of the word is more problematic. Does copying and pasting from Reddit and 4chan really constitute *producing* something new? The word *producer* carries significant cultural weight, akin to one who *produces* a film or a company that *produces* a new electronic gadget. In both cases, large amounts of capital, brainpower, and labor are expended with no guarantee of success. Posting a plagiarized meme on a loosely regulated subreddit does not belong in that same arena. Further, the

---

1   Gilbert, Nestor. "48 Reddit Statistics You Must Read: 2020/2021 Data Analysis & Market Share." *Financesonline.com*, 13 Aug. 2019, financesonline.com/reddit

2   Hecker, Rabea. *Are You Serious It Is Just a Joke? The Influence of Internet Memes on the Perception and Interpretation of Online Communication in Social Media.* 2020.

second definition of the word *produce*, as sourced from Dictionary. com, "to bring into existence by intellectual or creative ability," is not relevant to the common meme—a medium so fraught with piracy and stolen content. The poster who claims to *produce* the meme often didn't do anything at all, outside of executing a few right clicks. Therefore, the word *prosumer* grossly exaggerates the work and creativity required to meme. Warren Buffet's one sentence dismissal for gold as an investment applies to memes and their *prosumers*—"It doesn't produce anything."

ARE MEMES A VIABLE SOURCE OF JOURNALISM?

COMMONLY USED SOCIAL MEDIA sites serve as a de facto news source for many people. In reality, memes have been known to interfere with the real-life new cycle. One example comes from Croatian politics. Ivo Sanader served as the Croatia prime minister from 2003 through 2009, making him the longest serving prime minister of Croatia following their succession from Yugoslavia. In 2009, Sanader shocked the young nation when he opted to resign abruptly in 2009. In 2010, he was arrested on numerous charges of corruption, bribery, and abuses of power. Sanader remained in prison until 2015, when the Croatian Supreme Court dismissed the charges against him. Several months before his release, a Facebook group whose name translates to "Padre is coming back" formed, thus signifying the start of the meme cycle in Croatian politics.

Domagoj Bebic and Marija Volrevic, a political scientist from Croatia and a social scientist from Slovenia (respectively), began analyzing articles from the mainstream Croatian media cycle (from

October 1 until December 31 of 2015) against memes that were posted in the aforementioned Facebook meme group during that same time period.[1] Bebic and Volarevic began by evaluating the nearly 1,200 memes posted onto the Facebook page during the entirety of 2015. The memes were overwhelmingly positive toward the former prime minister, at 87 percent, with another 12 percent being neutral.

Here is where the duality of the results become interesting; Bebic and Volarevic then investigated over 100 new articles regarding Sanader during the three months following his release from prison on October 1. Three different Croatian news outlets were examined for the study. The news outlets reported on Sanader neutrally from 47 percent to 76 percent of the time. This shows the inherent disconnect between the two mediums. The memes, the heavy-handed medium devoid of context, seemed to have a substantial disconnect from the mainstream media articles (which mostly stated objective court facts and other relevant data in their vetted articles). In the microcosm of the *memeing* atmosphere, facts do not seem to reign supreme; meme posters form an opinion and clutch to it for dear life.

The researchers uncovered an even more troubling development as the pro-Sanader *memeing* appeared to have influenced 6 percent to 24 percent of Croatian mainstream media articles portraying the former head of state positively. Every single one of the articles that referred to Sanader as *Padre* were all positive. The word choice (and by proxy, the sentiment) of professional journalists was seemingly swayed by a bunch of memes. This is dangerous because, as previously discussed, memes lend themselves easily to misinformation and radicalization. When meme culture spills over into mainstream reporting, it will inevitably influence the ordinary citizen who would not seek out an extremist Facebook page for news.

---

1   Bebic, Domagoj, and Marija Volarevic. "Do Not Mess with a Meme: The Use of Viral Content in Communicating Politics." *Communication & Society*, vol. 313, no. 3, 2018.

As the field of journalism has adjusted to the modern digital landscape, more news outlets have used memes as a means to broadcast their stories. One of the leading newspapers in Brazil, *Folha de S. Paulo,* has not only used memes as a journalistic source but has also resorted to running memes in their print edition. Also in Brazil, the website MemeNews was launched, using consumer-submitted memes to accompany stories that their relatively small force of journalists cover.[2]

However, as the common meme-hosting websites have exploded in the twenty-first century and have morphed the internet from a broadcasting role into a participatory (although this book will later debunk the *participatory* part) one, something was lost along the way. As journalists have had to compete with common meme-hosting sites to get the eyes of the public onto their stories, they have been forced into creating digital video and graphical content to go along with their written prose. They have also needed to learn how to optimize the release of their stories to ensure maximum algorithmic exposure. It is highly unlikely that a lone journalist has all of these varied professional skillsets. While large and traditional broadcasters can rely on having a diversely skilled workforce to assemble these multimedia stories, smaller publications, such as MemeNews, have relied on a mostly volunteer workforce to assemble these elements together. Something was lost in the shift from traditional reporting sequence to the so-called *participatory media* (a term frequented by academia) of memes.

That something was the caveat that John Oliver extolled—*memes aren't facts.* This truth is not universally accepted in the world of modern journalism, and that is a crisis of the profession. Many claim that memes serve as a proverbial pulse of the public (despite their heavy tendency toward confirmation bias and dishonesty). However, they are fraught with half-truths and are light on context—

---

2   Bede, Isabelle. *Journalism Embeds Social Media Language: The Use of Internet Memes in Political News.* 2019.

yet truth and context are the bedrock of solid journalism. Memes are not a credible news source and should not be relied upon by the professional journalist, especially one attempting to present a faithful and balanced account of a story.

Thankfully, this trend has not gone unchecked. OhMyNews.com is a South Korean-based news agency that, like its Brazilian counterpart, MemeNews, relies on crowdsourcing for many of its stories. However, OhMyNews dedicates a small staff to perform the traditional and necessary gatekeeping process of fact-checking and editing prior to each story's release. Without this much-needed professional intervention, mass participatory models of reporting suffer from a lack of trust and objectivity (given the known echo chambers in the meme community). This robust quality control system is largely absent on common meme-hosting sites, and thus one needs to harbor serious suspicion before considering memes as a news source.

Another pitfall of using the common meme as a news source comes from the aforementioned crowdsourcing model. MemeNews is rumored to rely mostly on a volunteer staff to assemble their multimodal stories for consumption. MemeNews fails to reveal any information regarding their staffing practices or their sourcing integrity. The consumer is, therefore, left to speculate as to the source of their stories, or worse yet, take them at face value. As is the case with other large information and database hosting sites (such as Wikipedia, one of the world's leading nonprofit databases), these are uncompensated volunteers are often amateurs. The meme culture that America has created and hence exported globally is currently resulting in rampant economic exploitation.

MemeNews provides the memes as the headline of their stories, with the link to the story below. Once the user clicks that link, they are redirected to external news agencies such as *CNN, BBC, USA Today,* and other legitimate news agencies. MemeNews isn't even writing their own stories, merely using memes as a window dressing to entice viewers. This is intellectually lazy.

Adrienne Massanari, a communications professor at American University, examined the alleged participatory culture of Reddit and discussed the use of Reddit as a news source.[3] Under the guise of participatory *culture* (which we'll discuss later), subreddits rely on their users (often unpaid, untrained, and exploited amateurs) to post relevant news stories and biased memes. According to Massanri's (2015) examination, 62 percent of Reddit users (now totaling 330 million users) use Reddit as their primary news source. This has reduced the perceived need for journalists (as stated by Massanari). Substituting professional and ethical journalists for lowbrow *memelords* is dangerous to informed civil discourse. It also devalues the role of objective journalists. Memes are, due to their nature, created after the event in question has already happened and been digested by the original creator. Memes are not even a news source; they are an interpretation of the news. Memes are merely a reaction, rather than being a proactive story written by a journalist.

There are several good reasons that memes should not be used as a part of the professional journalist's toolbox. The Society of Professional Journalists publishes their Code of Ethics, which serves as a guideline for their members as to what constitutes ethical journalism. Let's examine the list of ethical standards and discuss how meme use runs counter to these stated practices.

The very first provision of the Code of Ethics (COE) is to "use original sources whenever possible."[4] *Memeing* will nearly always contradict this point. Memes, by their very design, are intended to be shared an endless number of times after their creation. Also, due to the copy and paste culture of many meme-hosting sites and their lax enforcement of stated conduct rules (take r/DankMemes, for example), it is usually difficult to pinpoint the original creator of the

---

3   Massanari, Adrienne. *Participatory Culture, Community, and Play: Learning from Reddit*. Peter Lang, 2015.
4   Society of Professional Journalists (SPJ). *Code of Ethics*. 2014, www.spj.org/ ethicscode.asp

meme. Thus, meme use by journalists violates the COE provisions of identifying sources clearly and never plagiarizing.

Another COE standard is to "Provide context." The brevity of the meme helps propel its journey across the meme atmosphere; however, this comes at the expense of context. Using MemeNews as an example, the memes are used as a secondary medium, and the context needs to be reinforced via the accompanying article. The article can be the viable stand-alone communication medium, while the meme cannot.

Further, COE states, "Support the open and civil exchange of views. Even views they find repugnant." Common meme-hosing sites, like Reddit, are known for taking the opposite approach. Many websites utilize algorithms that specifically gather and collect biases from each specific user. This creates an echo chamber, where each meme consumer is enveloped in memes that confirm their worldview and where little exchange of foreign ideas take place. Reddit takes this intellectual atrocity a step further by using the downvote mechanic. Redditors can downvote one another, and a post with enough downvotes gets pushed to the bottom of the feed before ultimately being hidden by default (meaning a user must proactively seek out the downvoted content instead of having it passively appear on the thread). This effectively serves as a form of censorship.

The COE goes on to advise journalists to "label advocacy and commentary." Memes are tools of commentary, as Isabelle Bede, a media scholar, noted.[5] Political and social memes are often created specifically for the purpose of taking an ideological stance on a divisive issue. Despite their brevity, many memes do not identify as such, leaning on the audience (willing or otherwise) to determine their own stances. For this reason and the several cited above, memes do not belong in the professional journalist's arsenal.

---

5   Bede, Isabelle. *Journalism Embeds Social Media Language: The Use of Internet Memes in Political News.* 2019.

AT FIRST GLANCE, memes seem like a great choice for the activist. Memes are created to take hardline stances on various political and social issues. It has been noted time and again that one effect of memes has been their ability to create a conversation, albeit one rife with confirmation bias. A term has entered our recent vocabulary that succinctly sums up much of the world of *memeing*-for-a-cause— *slacktivism*. The word, a portmanteau of *slacker* and *activism*, is defined by the Cambridge University Dictionary as "activity that uses the internet to support political or social causes in a way that does not need much effort."

Take for example the 2016 US presidential election. The campaign was marred by controversies associated with both Donald Trump and Hillary Clinton. Online, the memeosphere ran rampant, with each side's blindly loyal base of supporters churning out memes at a nearly breakneck pace. Many dubbed the 2016 election as the "Year of the Meme."[1] However, the election itself was hampered by a pitifully low voter turnout, where less than 60 percent of the eligible voting population decided to cast a ballot, the lowest since 2000.[2] Ultimately, the constant *memeing* stood for very little in terms of getting the impassioned fanbases to act on their beliefs in that one meaningful way.

Another prime example is the infamous Kony 2012 campaign that went viral near the beginning of the decade. A campaign surfaced on common meme-hosting websites regarding the heinous military practices of Ugandan warlord Joseph Kony. Kony was the head of an armed rebel faction based in Uganda and was known for committing a slew of war crimes, including the use of child soldiers. Suddenly, Western meme-hosting websites and their users were bombarded with messages, memes, and posts across several different platforms. The cultural fire had been set ablaze, and seemingly everyone was ready to "Stop Kony." However, the movement quickly died out and, at the time of this writing, Joseph Kony remains at large. After observing this phenomenon, one must stop and seriously ask *What did all these memes accomplish?*

Kien cited the Kony 2012 phenomenon as an example of how memes often perpetuate lies, or at the very least, misrepresentations of facts. The *Kony 2012* documentary had quickly amassed 110 million views. The Ugandan government prompted a response to the dubious claims made in the documentary, swiftly pointing out that the warlord had actually been driven out of the country several years before the video was made. The prime minister of the African country posted a response, aiming to correct misconceptions caused by the craze. Sadly, the Ugandan government's YouTube video had only amassed approximately 100,000 views. In essence, we tend to *fall in love at first meme.*

1   Noyes, Jillian. "Election 2016: The Year of the Meme." *The Odyssey Online,* 7 Nov. 2016, www.theodysseyonline.com/election-2016-year-meme
2   UC Santa Barbara. "Voter Turnout in Presidential Elections | The American Presidency Project." *Www.presidency.ucsb.edu,* 2021, www.presidency.ucsb.edu/statistics/data/voter-turnout-in-presidential-elections.

The meme posted above points out some weaknesses of memes as a medium, though also exemplifies some of those weaknesses (that is to say, it makes fun of memes yet reflects their weaknesses).[3] First, the use of humor trivializes the subject matter at hand and serves to belittle the suffering of others. The creator of this meme is likely a resident of a first world nation that is fortunate enough not to have to deal with problems like the one mentioned in this meme. Secondly, the bottom line of the text exhibits slacktivism, a problem endemic with the communication medium. Widespread criticism surfaced of Kony's actions and deservedly so; but there was also a widespread criticism of those who posted memes about the issue, and the above meme correctly identifies that the *memeing* had not actually accomplished anything. This meme succinctly sums up slacktivism, though it's also guilty of the same phenomenon.

It is important to note that this is not a stance against posting what one feels is right. It is, however, a condemnation of solely using online platforms to advance a cause with *no other action taken*. Specifically, if one is trying to convince outsiders to care about the

---

3   "Shallow Rave: When Memes Go to War; KONY 2012." *Shallow Rave*, 10 Mar. 2012,   shallowrave.blogspot.com/2012/03/when-memes-go-to-war-kony-2012. html. Accessed 2021.

cause being championed, it is worth asking, *Are memes really the most effective way to convey this message?* The answer is usually *No*, due to the tendency of memes to insulate a group from outsiders, rather than engage the world at large.

There are tangible ways to get involved in a cause. For example, one can start by finding a nonprofit organization that supports a cause they care about. There are several websites that monitor the fiscal responsibilities of nonprofit organizations, and rate them accordingly, as to how efficiently the organization in questions uses the funds received. For example, CharityWatch and CharityNavigator perform this service. If donating to charity is not your speed, try patronizing businesses that champion the cause you care about— whether that is through a portion of revenue going toward a trusted nonprofit or selling products that are manufactured and distributed in a way that impacts your cause (fair trade, organic, environmentally friendly, etc.). Even the antiquated (though still viable) method of writing a letter or calling your elected legislators are tangible forms of activism, that can and do impart real change upon society.

While memes claim to *raise awareness* of an issue, they often do so within profound limits. Flecha-Ortiz observed the use of meme discourse as a stress reliever during the COVID-19 lockdown of Puerto Rico. While memes may have been shared within the island of Puerto Rico, researchers found that traditional media (radio, print, television broadcasts, etc.) was more impactful on people's actions and responses to pandemic policies. So, of all the talk that meme proponents use of *raising awareness*, memes failed at this crucial task when they were needed most.

Raising awareness of an issue is not enough—it is merely the first step. If awareness is not followed by action, then we have to ask ourselves, *What are we accomplishing right now?* Take for example the viral "Ice Bucket Challenge" that took the nation by storm in the summer of 2014. The premise was simple enough—a participant would post a video of themselves dunking ice-cold water on their heads and, upon doing so, they would challenge some of their friends to do the same. The Ice Bucket Challenge became a powerful fundraising tool in the fight against ALS. What truly impacted the fight against ALS was the collective opening of America's wallets. *Memeing* isn't helping, even when it is at its most wholesome. Actions speak louder than memes.

An example of meme-based activism is the attempt to free Leonard Peltier from prison. Peltier was a leading member of the American Indian Movement (AIM), an organization that advocates for the rights of Native Americans. On June 26, 1975, two FBI agents were shot and killed during a tense standoff with AIM activists on the Pine Ridge Indian Reservation. Peltier was present at the event, although he denied killing the two federal employees. At trial, Peltier was found guilty and sentenced to two consecutive life sentences. With the advent of Facebook, the Leonard Peltier Support Group formed and began posting memes in support of the AIM activist, often demanding his release. Corinna Lenhardt, an ethnographer from the University of Munster in Germany, discussed this in her paper.[4]

Memes are rarely an effective form for encouraging activists to take action in real-life. Common meme-hosting sites are more content with fattening up their engagement statistics rather than mobilizing users. There are successful formulas for encouraging the viewer to take action in the real world, rather than just engaging Silicon Valley algorithms. Memes that include a very specific call to action and provide detailed, yet concise instructions on how to do so are an effective means of supporting the discussed cause. Unfortunately, these types of memes are rare, and this tactic is seldom used. An example is seen on the next page.

---

4 Lenhardt, Corinna. "'Free Peltier Now!' the Use of Internet Memes in American Indian Activism." *American Indian Culture and Research Journal*, vol. 40, no. 3, 1 Jan. 2016, pp. 67–84, 10.17953/aicrj.40.3.lenhardt.

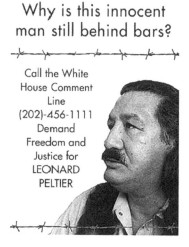

Why is this innocent
man still behind bars?

Call the White
House Comment
Line
(202)-456-1111
Demand
Freedom and
Justice for
LEONARD
PELTIER

This meme was posted to the Facebook page of the Leonard Peltier Support Group's (LPSG) on September 5, 2014. Here, the text conveys the group's message, pleading with the public to call the White House and advocate for the release of Peltier. This meme proved to be wildly successful (relative to other memes posted on the group's Facebook page). Supporters of the cause responded en masse by phoning the White House in support of clemency for their AIM comrade. However, even as Lenhardt herself has admitted, the odds of Peltier receiving clemency from then-President Barack Obama remained low. Even when memes prove to be successful in mobilizing their users, their results are questionable at best. At the time of this writing, Peltier is still incarcerated.

The LPSG, like many other meme groups who advocate for a cause, has also been guilty of dumbing down memes to suit their cause. The meme below depicts world-renowned physicist Albert Einstein. He was originally photographed writing a mathematics equation on a chalkboard, but the photo was altered. In place of the equation, it says, "Free Leonard Peltier" on the chalkboard; this is a meme that attempted to use a common humor vehicle (Einstein memes) to forcibly highlight a serious cause.

As Lenhardt acknowledges, this meme—rather than becoming a conduit for starting the grander conversation the LPSG had hoped for—instead became just a variation of the Albert Einstein meme; the meme had just become another punchline. Using memes as their communication medium of choice cheapened the cause they so deeply care about. Situations like this are why it's important to carefully consider the delivery of one's message by asking, *Does the medium match the intent of the message?*

There's an old adage among journalism and communication scholars that says, "the medium makes the message." Memes as a communication medium are flawed in the sense that they cheapen the message. Hecker demonstrated this by gathering a group of Facebook volunteers. Each volunteer was shown six Facebook posts, three with memes and three that were text only. They were asked to rate the seriousness of the message being conveyed. Relative to the text-only posts, the memes were rated higher in likability, though they rated lower in seriousness, as noted by Hecker's study.

Hecker's findings suggest that using memes to communicate one's ideals will cheapen the message in the eyes of viewers. This is a severe blow to sincere activists who are trying to spread their cause to potential recruits, since those prospects will likely disregard the posting as "just a meme." The other proverbial edge of this sword is that memes, because they are so easily dismissed, can also be used to spread controversial ideas.

Wee Yang Soh, an anthropologist at the University of Chicago, published a scholarly article detailing the supposedly pragmatic use of internet memes as a form of protest in Singapore.[5] Soh attempts to make a viable case as to why memes are an important medium for political dissent in the Asian city-state. Much like the studies referenced in the education chapter, Soh winds up making several damning concessions that actually weaken the case for memes.

Soh begins by stating that the Singaporean government censors many of the traditional media outlets within its borders, thus forcing many of the residents to resort to internet memes. (The point of this book is not to comment upon the validity of these claims, and hence will take Soh's claims of censorship at face value.) Soh states that while the Singaporean government can (and assuredly will) see the internet memes posted by its residents, the government is hesitant to police them because doing so would cause the government to lose favor with the West. Therein lies a valuable angle in the meme discourse that scholarly articles tend to neglect; one can only be a dissenting *meme-lord* in a developed first world country or a country that is trying to maintain good relations with the West. Given this, memes are inherently limited by geography to their effectiveness.

Meme proponents love to proclaim the medium's effectiveness for enacting widespread social and political change through creating a conversation (one that research shows is rife with confirmation bias and extremism). Nevertheless, assuming Soh has fairly represented the Singaporean government, Singapore remains a nation that maintains a strict control over media outlets. If memes supposedly can enact this alleged tidal wave of change that their proponents claim, then Singapore would not be able to govern in this manner. Said another way, the mere existence of this scholarly article debunks

5   Soh, Wee Yang. *Digital Protest in Singapore: The Pragmatics of Political Internet Memes*. Media, Culture and Society, 2020.

the claim that memes enact the scale of change their apologists advertise.

Soh goes on to acknowledge that memes are "not inherently meaningful or socially consequential on their own" and that memes "do not exist in isolation but achieve their effects in relation to other forms of social action." Memes *do not* exist in isolation because they inherently *cannot* exist in isolation, for the medium is reliant upon a shared understanding and often a long-form piece of communication to reinforce the meme, to both create context and (ideally) to clear up misconceptions. However, the observation that memes are not meaningful or socially consequential remains accurate.

Soh continues by pointing out that memes can still be posted by residents because the medium allows a certain degree of plausible deniability that the meme is a form of protest. The reasoning stated throughout the article is that the element of humor contained within the common meme allows it to be de-escalated from political protest to *just a joke*. That important distinction allows the meme (and its poster) to go unpunished. But, as we have seen earlier, meme-based discourse is also taken less seriously. Activism that isn't taken seriously is typically ineffective, hence why such self-dismissive tactics are not widely used for social change.

The *just a joke* narrative unsurprisingly backfires. While the intended effect is to have government censors dismiss the dissenting memes as *just a joke*, Soh concedes that the same mechanism has also applied to the average Singaporean. Humorous political memes are not taken seriously by their intended audience. And it's certainly a crippling blow (and a juxtaposition) when the intended audience finds the subject matter humorous as opposed to encouraging genuine change. This shows once more that memes are an inherently poor medium in the realm of social change.

Soh also touches upon the fact that the meme *poster* is not necessarily the meme's *creator*. Soh paints this in a positive light, stating that the meme being posted doesn't explicitly link the poster to the content within the meme, thus maintaining some sort of deniability of the dissent. The strongest attribute that memes possess is that the medium is so rife with piracy and plagiarism that it cannot reasonably be traced to its source, and therefore keeps the poster safe from retribution. If the Singaporean government can/does keep tabs on every resident's meme postings (as Soh claims), then it can surely perform a simple reverse image search. As an aside, *memes are so fraught with piracy, it's difficult to trace the source* is a poor defense of the medium. Secondly, the "degree of separation" argument Soh makes assumes that the poster has absolutely no agency whatsoever on the content posted on their page, and therefore cannot be held responsible for content posted on a page under their name. This claim does not hold up to scrutiny when one realizes that most common meme-hosting sites require a unique name and password in order to post (hence, debunking the "degree of separation" claim). Lastly, if the Singaporean government is as repressive as the author of the article claims (again, this book is not taking any pro- or anti-Singaporean stances), then it wouldn't be difficult for a government official to claim that posting a dissenting meme is the same thing as stating those views directly, and hence does not protect the poster from punishment.

Soh cites several news stories in which Singaporeans bemoan the medium's tendency to downplay the seriousness and urgency of the modern political climate of their country. Not only did the residents think that medium cheapened the message (which it does), but it is important to note that a segment of the population—those who memes are supposed to be helping—aren't being heard. The last thing the Singaporeans deserve is to have a medium forced on them when they collectively do not want it.

A 2017 study of the role of memes in political activism within three different former Eastern Bloc nations (Belarus, Ukraine, and Estonia) likewise demonstrates that internet memes are not an effective tool. Here, too, this was not the intention of the authors,

Karatozgianni, Miazhevich, and Dennisova. The paper set to portray internet memes as a viable means of alternative protest that could enact widespread, sweeping change while keeping the protesters safe from prosecution.

They start off by proclaiming that memes are allowed to exist in authoritarian countries because they can exist within the government's censorship apparatus. The governments in these countries allow the medium to exist to give the citizens an emotional outlet to prevent an insurrection. This so-called strength of the medium, in reality, reinforces the point that memes can only exist if an authoritarian government allows them to exist. This isn't a strength of the medium at all.

Karatozgianni, Miazhevich, and Dennisova further state that memes have limited use in creating real-world change, because memes can only flourish within our digital spaces, and they cannot create an offline community that enacts true change. This is another bold concession that another academic group of authors is forced to make about the medium, doubly so in a paper that hails the medium's virtues in enacting change.

Karatozgianni, Miazhevich, and Dennisova state that nations are still relevant in policy and legislation regarding the usage and subsequent spread of digital information. This runs contrary to the common pro-meme argument of *memes are borderless and can unite those of different cultures.* Memes are still very much limited by geography and do not serve to level any inherently unequal playing field.

Karatozgianni, Miazhevich, and Dennisova further mused on the definitions of "political participation" and "mobilization." Political participation has been cheapened to "any manifestation of one's political views, persuasion of others and politicized artistic contributions to the digital sphere." By removing the onus of *political participation* from any real-world responsibilities whatsoever, these authors have essentially lowered the bar so that any act is *political participation.* Further, the definition of *mobilization* defined in this scholarly article is the "deliberative activation of social networks to diffuse political information and influence the opinions of others" and continues, "it does not call for offline action of establishment of organizations or structures." Using this weakened definition, Big Academia can feel free to claim that memes encourage this so-called political participation. Meme proponents and Big Academia have made numerous concessions in their arguments. The only way they can convince the public of the efficacy of memes is by changing definitions. In order to shoehorn their narrative down our collective throats (even as evidence to the contrary mounts), meme proponents are willing to childishly change the rules of a game when they are losing and/or openly lying to the world about their dying medium.

Karatozgianni, Miazhevich, and Dennisova interviewed Chris Atton, a media professor at Edinburgh Napier University in the UK. Atton states that activist-minded meme posts are "actions in their own right rather than a platform for broadcasting about real change." This is a blatant confession from the pro-meme community that *memeing* doesn't bring about any lasting change, nor does it even require real effort beyond posting a meme. *Memeing,* even in the circles that have the most vested intellectual interests to extol their virtues, are a weak medium and do little more than encourage slacktivism.

Karatozgianni, Miazhevich, and Dennisova go on to discuss the meme-based political discourse in Belarus, a former piece of the USSR. The authors profess that new alternative media such as internet memes "remain mostly sporadic one-off localized initiatives." This sentence is an important concession to the *memes encourage offline activism* narrative, since it serves as a blatant confession that memes in fact do not cause sweeping change. The authors disclose that, in Belarus, memes do not typically migrate into offline activism. Rather, meme-based political dissent starts and ends online. This is a textbook case of slacktivism. While it is important to keep in mind the authoritarian nature of the governments of the Soh article and in the Karatozgianni, Miazhevich, and Dennisova discussion, the pattern of memes not encouraging real-world change has stayed consistent in nations around the world, Eastern or Western, free or tyrannical.

Karatozgianni, Miazhevich, and Dennisova discuss that feel-good online activism does little to change anything in the real world. This matter holds true in many of the cases we have examined in this chapter, in former Eastern bloc nations, the United States, Singapore, and more. Memes do not create offline activism, but merely encourage slacktivism. Lastly, this study ends on discussing Estonia. Estonia is a former USSR state that gained independence after the fall of the Soviet Union. However, Estonia is a much freer society than the other nations discussed in the article. The article praises Estonia for their radical transparency and their commitment to free speech. This is definitely something to celebrate (as this book absolutely endorses free speech), But the study makes no mention of what role memes had to play in that. While this is speculation, based on the evidence presented thus far, it's fair to speculate that memes played absolutely no part in Estonia's free-speech status at all. For an article that claims to examine the role of memes in these societies, the article didn't examine Estonian memes at all. This omission is something to keep in mind when reviewing a pro-meme academic narrative.

In their study of the use of memes during the COVID-19 pandemic in Israel, Kertcher and Turin observed that Israel was in the midst of a political crisis when the pandemic struck. The crisis revolved around a prolonged government shutdown, as well as the prime minister facing corruption charges. However, they observed that only 15 of the 436 memes addressed the issue of the corruption charges. The overwhelming majority of the memes did not address the government shutdown either. The focus of the memes was humor surrounding the COVID-19 pandemic. This shows that memes don't mobilize their audiences to take action in real-life, and they aren't equipped to deal with the serious topics of the societies they intrude on. Another hypothesis is that the digitally powerful elite have used memes to easily manipulate public discourse away from political underachievement and toward the Chinese being the alleged source of the pandemic. From addressing Israeli political dysfunction—and by proxy, encouraging any real-world democratic action—to encouraging adherence to COVID guidelines, memes have failed Israel.

Soh, in his 2020 article, goes so far as to say that memes are a "pathological symptom of the failings of modern society." I couldn't agree more!

DO MEMES STAY RELEVANT?

ONE REASON THAT MEMES are copied and pasted and proliferated so quickly is because they have no choice. Lilian Weng, a computer scientist from Indiana University, published a 2012 study that demonstrated how memes must compete for survival in our digital landscape.[1] Memes have a half-life, and a surprisingly short one at that. This is why many meme posters are forced to pump out as many memes as possible, sacrificing quality in the process. The figure below represents the meme half-life of the archaic medium. This finding was also upheld through another study performed in Puerto Rico by Flecha-Ortiz during COVID-19 lockdowns. Memes were observed in the final hypothesis of the study (which was ultimately supported) to quickly lose relevance after providing an ideological convergence on an issue to a specific in-group. Memes as

---

1 Weng, L., et al. "Competition among Memes in a World with Limited Attention." *Scientific Reports*, vol. 2, no. 1, 29 Mar. 2012, 10.1038/srep00335. Accessed 14 Nov. 2021.

a communication medium tend to quickly die off after their collective use has been fulfilled. Flecha-Oritz also found that, in order for memes to spread, a particular meme must live in the memory of the user for a certain amount of time, providing an inherent limit on the longevity of memes (and with all due respect to Richard Dawkins, that is not *the rest of the recipient's life*). However, these varying interpretations quickly converge into a single opinion (as much of the scientific literature cited in this book has demonstrated), thereby resulting in an echo chamber. So, in order for memes as a whole to ensure survival, mass production must be deployed, not unlike pollen from trees (except memes aren't vital to our cultural ecosystem).

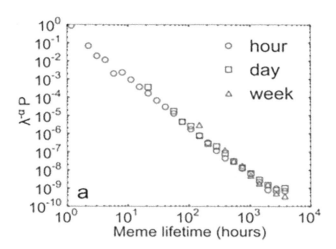

This graph, from the Weng study, measures the probability of distribution on the vertical axis, that is to say, the likelihood that a particular meme will be distributed on Twitter. The horizontal axis is the lifetime of the meme, measured in hours. As we can see, there is a clearly negative relationship between time and chances of distribution. At merely ten hours, a meme stands only 0.001 percent of its original opportunity to be shared. At one hundred hours (a little more than four days), that same meme stands only a 0.00001 percent of its original likelihood of being reposted. This is one reason the internet is constantly bombarded with memes; it is their only means of survival.

This distribution is not unlike what is observed in nature—with one important difference. Often, an organism will produce far more offspring than will realistically survive, to ensure genetic survival against the harsh realities of their ecosystem. In much the same way, a Redditor is forced to post as many memes as possible to bolster their chances that, statistically speaking, at least one of their posts will make the front page of the meme-hosting site.

However, memes are dissimilar from gametic cells in a very key way; reaching the front page of Reddit means nothing in the long run, for Redditt's front page washes anew every twenty-four hours. By contrast, once a sperm cell fertilizes an egg, a new offspring begins to form, thus creating a life that can hopefully survive until maturity. This life creates a competent, sentient being that can interact with the world around it.

Similar to the Weng study, Christian Bauckhage performed a time-based statistical analysis on 150 memes across various platforms. Like the Weng study, Bauckhage found that memes often have a quick stratospheric rise in popularity, followed by a sharp decline back to irrelevancy. Another wrinkle that Bauckhage found is that memes often rely on novelty to survive. This isn't surprising, given its short shelf life.

The thing is, much like last week's garbage, memes don't just magically disappear; they have to end up somewhere after their (sometimes literal) fifteen seconds of fame. The internet is a place of permanence, where one's posts are rarely, if ever, truly deleted. Time and again we have seen athletes and entertainers rise to fame only to be marred by controversy when a pre-fame tweet or other posting is discovered. However, it seems that these meme posters don't seem to realize that. This was predicted in 2001's hit PlayStation2 game *Metal Gear Solid 2*, in the final speech delivered by Colonel Roy Campbell, the main character's commanding officer, when he mentions that digital waste is accumulating at an alarming rate.

However, digital waste can also serve useful to companies and campaigns alike. As Bauckhage mentions, memes are a research interest in the realm of data mining and analysis. This study was performed in 2011, well before the Cambridge Analytica scandals revolving around the UK's Brexit decision and the 2016 US election, where Cambridge Analytica performed vast data mining to create targeted advertisements on behalf of the campaigns who hired them. The purpose of this book is not to demonize Cambridge Analytica; after all, they are a for-profit company that acted within the bounds of the law to mine public or legally purchased data (data that was voluntarily given to Facebook by platform users). Cambridge Analytica also has a fiduciary responsibility to customers and shareholders. The point is that memes provide excellent fodder for those who wish to data mine and monitor insulated communities. Bauckhage also speculated in the discussion section of his paper that the analysis he performed can inspire future work that involves using the time-based statistical analysis to predict future meme usage. From an academic standpoint, this does not inherently seem like a bad thing. But those who have an agenda or wish to manipulate discourse within isolated communities can very easily use memes to alter future discussions revolving around a topic, akin to the final AI speech from the classic video game *Metal Gear Solid 2.*

The half-life of memes limits their usefulness in the political arena. A common defense mounted by rabid meme posters is that memes can start a discussion on an important topic, which then turns into engaging civil discourse. As we've seen with the Weng study, memes typically live a very short life in the realm of cultural relevance. Therefore, their opportunity to spark this supposedly great exchange of ideas is markedly brief. This requires the poster and their ideological cohorts to post as many memes as possible in order to boost their chances of creating this dialog.

Memes are also inherently limited in the dissent political arena, due to their underappreciated geographical constraints. For example, if country A is ruled by an authoritarian government whose regime tightly controls and censors the internet in their jurisdiction, then memes are unlikely to be an effective means of political dissent.[2] Memes are limited to the developed democratic nations with potential influence. In the case of the citizens of hypothetical country A, their meme-touting saviors must come from the outside to post their grievances. With the reliance upon foreign *meme-lords*, who are often unfamiliar with the cultural and political contexts of country A, the memes are likely missing a lot of important details and context that would have made the meme relevant. It's inherently much easier to be a dissenting *meme-lord* in a developed democratic country.

Other arguments for the longevity of memes fall short. As noted by Karatozgianni, Miazhevich, and Dennisova from the previous chapter, some scholars liken memes to the folklore of digital crowds. This is a mighty comparison that does not hold up to scrutiny. Folklore implies that the stories, thoughts, culture, and messages conveyed last for generations, even centuries. As we know, memes have a considerably shorter shelf life. To compare the common meme to folklore is overly optimistic at best, and downright arrogant or deceitful at worst.

Dawkins goes so far as to claim that memes travel down generations. However, as the Weng study demonstrated, memes as a communication medium cannot survive for multiple generations if they cannot even survive for hours, let alone weeks. Big Academia and Silicon Valley frequently try to prop up their failed medium, often ascribing feats to memes that are demonstrably impossible, similar to *Chuck Norris facts*. However, the major difference between memes and *Chuck Norris facts* is that *Chuck Norris facts* realized they were tongue-in-cheek the entire time. It is the hope of this book

---

2 Denisova, Anastasia. *Political Memes as Tools of Dissent and Alternative Digital Activism in the Russian-Language Twitter.* 2016.

that memes follow the way of the *Chuck Norris facts* trend and finally cease.

Theoretically, every meme has a chance at exponential distribution, as noted by Kertcher and Turin in their 2020 Israeli study. However, the average meme will never experience this level of popular sharing. Memes die a forgettable death shortly after they are posted. Amanda du Perez and Elaine Lombard discuss Richard Dawkins's meme hypothesis, specifically including longevity as one of the characteristics that truly define a meme. However, memes are a hyper-disposable communication medium that rarely survive more than a handful of hours, let alone one week. Therefore, if the pro-meme community insists on using this as a yardstick for what constitutes a viable meme, then the number of viable memes drops considerably.

MEMES & EXTREMISM

MEMES ARE USEFUL for many nefarious purposes. They are an ideal medium for spreading controversial, even hateful, ideas. Memes with an extremist slant can have a desensitizing effect on the viewer, especially those within an in-group, per Hecker's findings. Memes are also an effective tool in the hands of unsavory individuals or groups to manipulate the masses, because visual rhetoric is known to better sway one's emotions. The desensitizing effect and the illusory truth effect discussed earlier are a dangerous combination because they can cause the viewer of these memes to internalize the extremist ideas conveyed.

Upon doing due diligence and research for this book, I was tipped off that a common source of memes were secret Facebook groups. These groups, which a user must be invited to enter after being vetted by a gatekeeper, have the option of not being searchable or visible to the general public. (Disclaimer: we won't be referring to closed off, yet still searchable *private* Facebook groups, only the *secret* Facebook groups that remain hidden altogether.) In an *NBC* article, a spokesperson for Facebook claimed there were "tens of millions" of Facebook groups, though declined to say how many secret groups there were.[1] Information has been sparse as to the exact number of secret groups the platform is hiding from the public. However, the same *NBC* article managed to reveal that the combined number of members in the radicalized, Republican conspiracy group, Qanon, was approximately three million users. Another *NBC* article cites that there are 1.86 billion Facebook users in total.[2] This mathematically works out to about 0.05 percent of users on the popular meme-hosting site having been radicalized. For the sake of argument, let's assume that radicalized Democrats are also *memeing* in the dark to the same degree (this book does not take ideological stances). With that assumption in place, we can now figure 0.1 percent of users have been radicalized. With this estimate in place, we can deduce that any page with at least 1,000 members has one radical lurking in the shadows.

Without any visibility or accountability from Facebook (the meme-hosting site refuses to publicly comment on steps taken to combat radicalization), one is only left to speculate. If memes are common in these groups as we are led to believe, and these secret groups are radicalized, then we can deduce that there is indeed some relationship between memes and radicalization. The only way to know for sure would be if the meme-hosting site chose to shed some light on the secret groups. However, the Silicon Valley Elite have declined to do so.

Of course, it would be utterly naive to say that the United States

---

1   Sen, Ari, and Brandy Zadrozny. "QAnon Groups Have Millions of Members on Facebook, Documents Show." *NBC News*, 2020, www.nbcnews.com/tech/ tech-news/qanon-groups-have-millions-members-facebook-documents-show-n1236317.

2   Newcomb, Alyssa. "Who Is Policing Facebook's Secret Groups?" *NBC News*, 2017, www.nbcnews.com/tech/social-media/who-s-policing-facebook-s-secret-groups-n729856. Accessed 26 Nov. 2021.

is the only nation plagued by meme radicals. A group of Finnish researchers observed and analyzed several years' worth of memes posted to two extremist groups based in Finland.[3] The two groups, known as Finland First and Soldiers of Odin, are in-person groups that have used memes to spread their messages of anti-multiculturalism online. The study's introduction page mentions that memes have become a tool of extremists due to their ease of sharing, concise messages (though intentionally lacking context), and visual format. One such meme, pictured below, was posted by Finland First.

Figure 1. Posted: 23 November 2015 (Were they racists?).
Source. Retrieved from: https://facebook.com/Suomi-Ensin

This image, sourced from the Finnish study mentioned above, illustrates some key issues with the meme format. The image depicts a group of soldiers laying in the snow. The Finnish text leads the viewer to believe that these soldiers are Finnish, regardless of any supporting evidence to that claim. There is nothing in this image that would convince the layperson (which is who this meme is presumably marketed for) that these soldiers are even from Finland. Furthermore, there is no evidence of when this photo was even taken. Such context would have been helpful in confirming whether these

3   Hakoköngäs, Eemeli, et al. "Persuasion through Bitter Humor: Multimodal Discourse Analysis of Rhetoric in Internet Memes of Two Far-Right Groups in Finland." *Social Media + Society*, vol. 6, no. 2, Apr. 2020, p. 205630512092157, 10.1177/2056305120921575.

soldiers were Finnish and why Finland was fighting in this unlabeled war. For the sake of argument, let's assume that these soldiers are indeed Finnish, as the meme proclaims. Finland, during WWII, fought against the Soviet Union in 1939 and provided safe haven for German troops during the German invasion of the Soviet Union in1941.[4] Such historical context is stripped away, likely on purpose, to paint the Finnish in a more positive light. This meme also suffers from many of the graphic design faults previously discussed; the white text is a poor color choice against a snowy backdrop and an overcast sky. A different color choice would have helped the text's visibility, and hence its believability. The lack of context given by the image (likely intentional, given its radicalized source) is a poor support for this meme. The caption—"Were they racists?"— ignites a perfectly acceptable response gathered by the given evidence—"I don't know. *We're they?* They did help Nazi's, after all." It's as if the creator of this meme (*assuming* they were from this Finland First group, though that is quite a stretch to say) either did not put a lot of effort into constructing this meme or blatantly chose to omit key historical facts to preserve a narrative.

If we look at the basic model of the meme-hosting site, we can see why social media companies are hesitant to do anything about the meme-based destruction of our social discourse. Shedding light on hateful meme groups would cause many users to delete their accounts. But sites like Facebook need as many users as possible to post as much as possible in order to generate data points for the all-important algorithm. Without the radicals and *shitposters*, the algorithm would not be able to scrounge data on the user base as quickly or efficiently. Therefore, Zuckerberg and other Silicon Valley Elites are incentivized to turn a blind eye to the meme-based radicalization of our society.

While it is not this book's stance to say that memes radicalized the

---

4 Augustyn, Adam. "Finland - Nordic Cooperation | Britannica." *Www. britannica.com*, 2010, www.britannica.com/place/Finland/Nordic-cooperation

very first members of these extreme groups, the brief communication medium has undeniably played a hand in recruiting others to the base of radicals online (though, as we learned in the activism chapter, memes rarely encourage real-world action, thankfully). Without the luxury of context, memes lend themselves very nicely to those looking to recruit others to their cause (whether they are extreme left or extreme right).

However, it *is* this book's stance that memes have served as a lightning rod in ways that other communication mediums haven't. Longer and more comprehensive methods of communication, such as books, magazine articles, and long-form essays, have not been used as frequently by those looking to radicalize new members. The longer forms of communicating one's extremist views leave the author more susceptible to criticism, and they risk exposing oneself and their lack of expertise. Memes, designed by their very nature to be quick and punchy enough to omit context, inherently do not have this issue.

Sadly, meme radicalization can have real-life consequences. On March 15, 2019, a man named Brenton Tarrant opened fire on two separate mosques in New Zealand. Tarrant subsequently posted on 8chan (a common meme-hosting site) that it was "time to stop *shitposting* and time to make a real effort." Tarrant was seemingly fed up with the slacktivism behind his extremist cause. Similar mass shootings took place in the wake of Tarrant's attack, likewise followed by posting their far-right ideology on 8chan. Calls (or dares) to commit mass violence were posted on the /pol/ (shorthand for politically incorrect) board of 8chan, resulting in the gamification of such events. Memes have become a communication tool of the far right and a central part of its culture. Memes have also served to glorify violence and other extremism and trivialize the suffering of outsiders, as noted by Crawford, Keen, and Suarez-Tangil. Thankfully, incidents of memes encouraging real-word action remans a rare occurrence.

Crawford, Keen, and Suarez-Tangil performed a study analyzing the memes posted on the /pol/ board of 8chan and cross examined them with other related sites, such as 4chan. Unsurprisingly, they found that the memes shared on these sites often contain racist or other discriminatory undertones under the guise of humor. More interestingly, they analyzed the memes that were copied and pasted between Chan sites. They found that the longer and less thematic texts (or non-thematic posts) were the ones that crossed sites more frequently. The researchers hypothesized that this was because different subcultures existed between each site (for example, 8chan's /pol/ board would have a slightly different culture than 4chan's /pol/ board). The long-form text posts served to create context for the regular users of the hosting board (context, as we have seen, does not come naturally to memes). They observed that the long-form text postings were serious in nature and did not attempt to be humorous. However, they also tracked the overall popularity of memes observed and found that the simple format memes (large blocky text, still image, etc.) were still the most popular overall.

The communities on the Chan boards were also observed to have formed their own meme discourse within their boards. The anonymous users fostered their own dialog of inside jokes that seemed foreign to the outsider, and their meme content reflected this exclusivity. This is one reason that long-based text posts were common when communication occurred across the Chan sites; each site had formed its own in-group. Memes in their traditional format are ill-equipped for building bridges, like this task required, so users relied on detailed text posts that gave proper explanations and context (something memes are incapable of). Even in the extreme examples portrayed in the Crawford, Keen, and Suarez-Tangil study, memes serve mostly to communicate with an insulated in-group, to the exclusion of outsiders. This notion has been supported in much of the other scientific literature cited in this book.

When meme posters do attempt to sway general public

opinion, they often do so by trying to lodge memes into political conversations, in the hopes of turning them political, a proposed *benefit* suggested by Karatozgianni, Miazhevich, and Dennisova. This often occurred with an audience of otherwise disinterested people. Memes are knowingly being used in an attempt to pull people into their ideology. Thankfully, as evidenced in many pages of this book, memes are often unsuccessful at radicalizing others off-line.

Yus further explores the impact that sharing memes has on the user on a subconscious level. The paper discusses how memes are shared within an in-group to reinforce the cultural norms of the group and to communicate their acceptance of these norms (whether benign or hateful, as we saw in the 4chan study). When member A shares a meme with member B, member A is subconsciously communicating, *You're a member of my group.* Eventually, humans will, consciously or otherwise, conform to the cultural norms of the group they belong to. Kien explores this concept as well, adding that meme users will edit content to represent their sense of self. Humans conform to the culture of the larger group of which they belong to. Those who post radical memes must realize that, in a very real way, *you are what you meme.*

Yus also discusses how the grammar and word choice of the meme in question ultimately cannot be fully divorced from its creator in a cultural sense. Memes largely exist to reinforce the norms of an isolated in-group (rather than the "bridge building" hypothesis the common Redditor will claim), seemingly as if to prod conformity among members of the in-group, forming a hive mind. Memes ultimately do not serve as a means to build bridges; they serve as a means to police one's cohorts. Yus also found that during the use of memes within an in-group, the memes that reinforced existing cultural norms within the group were lauded more than those who created something new from outside norms. That is to say that memes that supported the echo chamber were praised while novel ideas were not. This shows another flaw of memes; it is only useful for broadcasting a message that recipient is guaranteed to agree with it. These trends are even more troublesome when observed in relation to the radicalized 4chan and 8chan boards referenced in the Crawford study.

Given that each subreddit and 4chan board has a slightly different set of cultural norms regarding their meme communication, it is important to view this in the context of communication that can cross the gaps. Memes that jump across these boundaries are often altered along the way, as noted earlier. With each altering, their original message is lost more and more, to the point that the original intent is no longer recognizable.

Huntington also discusses memes as a form of visual rhetoric and the role that rhetoric plays in persuasion. The term *rhetoric* is an interesting word choice by Huntington, as it implies that memes are not factual—whether wholly or just partially. Putting aside the implicit concession made by Huntington's word choice, memes are often used by those with an ideological slant. The link between persuasion and rhetoric is a troubling one because meme creators can use this tactic to convert others to their cause. Thankfully, memes are an ineffective means of persuasion since they are not taken seriously and typically signal to those already in the in-group.

Women as a population have also suffered at the hands of internet memes. Jessica Drackett, a psychologist at Leeds Beckett University in the UK, performed a qualitative analysis on 250 internet memes sourced from various corners of the meme-verse.[5] One observation was that memes have been used, under the guise of humor, to spread sexism and misogyny through the dark corners of the internet. One of the points also touched upon in their paper is that memes were often used not only to reinforce group norms but also to police those who rebel against those norms. Since that mob mentality is typically

---

5 Drackett, Jessica, et al. "Old Jokes, New Media – Online Sexism and Constructions of Gender in Internet Memes." *Feminism & Psychology*, vol. 28, no. 1, Feb. 2018, pp. 109–127, 10.1177/0959353517727560.

not conducive to fair and equitable treatment, it can be asserted that memes are not a tool used for the benefit of those with dissenting (in this case, humanist) views.

Drackett and her team also discussed the inclusion of irony, emojis, and unnecessary suffixes like "LOL." These linguistic devices are used as a veil to spread hateful anti-woman messages. The paper explores various theories of humor and touches upon the superiority theory. As the name implies, this type of humor derives laughter from the misfortune or suffering of another class of people (in this study, women).

The Drackett paper also explores the concept of rebellious and disciplinary humor, specifically through internet memes. Rebellious humor (used to subvert the status quo) is also used in the medium; however, it is often used as a heavy-handed "subversion" of political correctness itself. This alleged subversion of political correctness is little more than an excuse to aim discrimination and hatred toward a group of people. At this point, rebellious humor goes full circle and once again becomes disciplinary. Such a feat would be more difficult to accomplish through longer forms of communication because memes are a visual medium that relies on the image hitting the viewer first, contrasted against a long-form text post where it is more difficult to hide one's intentions.

Amanda du Perez and Elaine Lombard further discuss the role of memes and how they survive. "Only those memes that are suited to their socio-cultural environment are likely to spread," according to the article. This is an important distinction because this shows that memes are likely to create a hive mind in which memes posted reinforce narratives. In the context we are discussing, memes with extremist views are more likely to spread in these groups versus benign ones. This certainly does not bode well for society at large.

As the Drackett study demonstrates, memes often use humor to defend or justify extremist views. Labeling the misfortune of others as *just a joke* is often used as a crutch to excuse the poor behavior of the meme posters. The study wraps up by discussing how social networking (and by extension, the common meme-hosting websites) become a means of constructing one's identity in the twenty-first century. Again, in a very real way, *you are what you meme.*

ARE MEMES PARTICIPATORY?

NUMEROUS ACADEMIC STUDIES have lauded memes as a sign of new media, one called *participatory culture*. However, the term *participatory culture* is often not defined. One must ask, *Participatory for whom?* Outside of self-selecting 4chan boards and subreddits, Facebook and Instagram are often glaringly public venues, where one's postings are viewable by their entire social circle, even those who aren't of the same cultural slant as the poster. For example, posting a meme on Facebook as a status means that even Grandma will see it. From the hypothetical grandmother's point of view, this wasn't participatory. Or again, think of the group texting thread one may be a part of; one member of the chat litters the chat with a meme, but odds are that at least one member of the thread wishes it wasn't so inclusive. Kien mentions that memes are often forced upon the masses "like a surprised bystander being tossed a hot potato." The implication is that memes are a communication medium of coercion rather than of mutual consent. So, we must ask ourselves, *Are these cultures truly participatory?* Memes are seldom a means of civil dialog; they are, instead, a megaphone. Megaphones are participatory only to the one holding the device, at the expense of all others around them.

Memes are a means of signaling a sense of belonging to other members of the in-group. But it is pertinent to ask, *Why, after becoming a member of this supposedly coveted in-group, do members feel the need to constantly assure others of their membership?* Are subreddits and 4chan boards so quick to cast off members for not *shitposting* regularly? How weak are these community bonds where the members have to constantly affirm themselves in the form of regurgitating a meme? We've explored earlier in this book that *memeing* tends to be a self-isolating hobby, with respect to the real world. So, we must question the strength of these meme-based ties.

Various subreddits have requirements of a minimum Reddit *karma* score in order to post. *Karma* on the meme-hosting website refers to the difference between upvoted content and downvoted content, with a higher score indicating a higher amount of upvoted content. The reason given is to prevent supposed trolls and *shitposters*, and to acclimate new users to the site. However, as this book has established, a primary focus of memes is to reinforce/police to other members of the in-group and enhance conformance to the cultural norms and beliefs systems of the group. Understanding this, it can be reasoned that karma requirements act as an ideological gatekeeper of the subreddit. Therefore, meme posters and those who visit that platform cannot seriously claim that the medium enables a participatory culture, due to this exclusionary effect.

Redditt's algorithm weighs newer posts and their initial acquired karma more than karma acquired down the line. This propels popular new posts to the top of their hosting subreddit, in a phenomenon known as the "Knights of the New." Then the human moderators weigh their judgment upon the posting within their realm of the website, often against vague criteria of their subreddits, which usually devolves into using their own slanted discretion. Reddit has several levels of gatekeepers; however, these gatekeepers are merely *ideological* (as opposed to the accurate or grammatical gatekeepers of the traditional media world). This ideological gatekeeping is another strike to those who claim that memes are a conduit of a participatory culture.

Karatozgianni, Miazhevich, and Dennisova wrote a study examined earlier in this book that regarded the use of memes as political dissent in three former Soviet bloc counties. One of the supposed benefits of using memes in this context is that the memes were able to inject themselves into mundane and nonpolitical conversations. One of the goals of this medium is to have memes infiltrate and pull uninvolved scrollers into a discussion regarding the meme's content. The fact that this is a secondary goal of memes further demonstrate how memes are not participatory. If something is forced upon someone who is not involved and does not wish to be involved, then we cannot deem that to be participatory. Perhaps a more accurate term is *coercive* or *invasive* media.

Memes are not a means of participatory media, no matter how much Big Academia—and their previously established weakened definitions—wants them to be. But memes have provided a chillingly accurate fodder to a twenty-one-year-old prediction regarding a dystopian society.

THE END OF THE CLASSIC 2001 video game *Metal Gear Solid 2*, features a twelve-minute conversation between the main character, an agent named Raiden, and the artificial intelligence (AI) posing as both his superior officer and his girlfriend.[1] The AI is part of a fictional shadow government that seeks to control the digital flow of information in order to control the thoughts and behaviors of the public. The speech given by the AI to Raiden has proved chillingly predictive of the meme-based decay of our society.

To start off, the AI states, "Trivial information is accumulating every second, preserved in all of its triteness. Never fading, always accessible." The situation, the AI adds, leads to "all of this junk data preserved in an unfiltered state, growing at an alarming rate." These words contain several gems of predictive truth buried within. As previously discussed, the spread of memes occurs at a rapid pace, which is required by their alarmingly short period of proliferation.

---

1   Kojima, Hideo. *Metal Gear Solid 2: Sons of Liberty*. Video Game.

That is to say, memes have a very short window in which they are shared. Many memes are apolitical and asocial, meaning that their content matter is typically tied to the quippy and lowbrow humor required by the two-line medium. The short half-life of the common meme typically means that the average meme poster (not to mention meme posters as an aggregate whole) needs to unleash a cascade of meme-based filth upon the masses, to ensure at least one of them reaches internet popularity. It's important to note that releasing a bunch of memes does not inherently boost the chances that a specific meme will become popular, as these are separate phenomena. Releasing more memes merely boosts the chances that *one* of them becomes popular. As previously mentioned, the internet acts as a graveyard of permanence, where after the hours-long decay of memes occurs, memes do not magically disappear—they are nearly always searchable. An underrated consequence of this meme pollution is that it opens the door for data-mining algorithms, not unlike the ones used by Cambridge Analytica, to acquire and analyze meme data and plot it against time. These algorithms then dictate what advertisements or stories appear.

In *Metal Gear Solid 2, the* AI continues to enlighten Raiden about junk data, which is filled with, "rumors about petty issues, misinterpretations, slander." With lackadaisical content screening by common meme-hosting websites, memes are left to proliferate misleading or downright false information (thanks to the illusory truth effect), as well as information that only appeals to a niche group. As we observed in the University of Michigan study, there needs to be is a largely visible statement used to boost the memes' believability. Believable memes have a much higher chance of being shared, regardless of any factual truth contained within.

For one moment, let's suspend any idea of intentional lies on the part of meme creators and argue that they sincerely believe the content they are spewing. Their cultural viewpoints can, and often do, differ from that of an objective viewer, and this can cause the holders of two differing viewpoints to interpret the same event in two completely different ways. However, with the inherent flaws of memes, there is no room to add the context of how one arrived at that interpretation. Memes are an inherently flawed medium for communicating one's ideas, though the rub seems to be that memes have been increasingly used for that purpose, much to the social unraveling of our nation and other developed nations around the world.

The AI continues to educate Raiden, lamenting the growth of junk digital data by saying, "It will only slow down social progress." At present, we seem to be frustratingly trapped in our social yesteryears, still strife with discrimination. These websites perpetuate hateful discourse, marred with radicalization. Memes are commonly used by hate groups, due to their spreadability and believability (independent of any truth). Rather than banding together, America is still very much a divided nation, in so small part due to the meme-based discourse circling within online pockets of hate.

Next, the AI drops another intellectual gem on our protagonist: "The digital society furthers human flaws and selectively rewards development of convenient half-truths. . . The untested truths spun by different interests continue to churn and accumulate in the sandbox of political correctness and value systems." It is easy for those with an agenda to create memes that may seem benign but have an ideological slant (rarely a positive one). We see this with memes that may not be outright lies, but which are loaded with confirmation bias. Memes in social and political groups tend not to challenge the status quo of their existing groups but rather to reinforce current group beliefs. *But this can happen with any communication medium*, the meme-defender would say, to which the response is simply, *Memes are often a tool that can spread such views, as they are easily digestible to the viewer.*

The AI enlightens Raiden about a context in which "Everyone withdraws into their own small gated community, afraid of a larger forum. They stay inside their little ponds." This line also reinforces the confirmation bias mentioned in the previous paragraph. However, the reference to being "afraid of a larger forum" part is interesting, because it shines a spotlight on the dark corners of the internet, the secret Facebook groups where hateful memes are peddled. As already discussed, Facebook will not make these groups (or their members) known under the veil of protecting privacy. (Although, in what other respect is protecting privacy part of Facebook's business model?) This has created the "little pond" phenomenon described by the AI from *Metal Gear Solid 2*. Even benign meme use in these groups can stunt the social skills of its users. After all, why be afraid of the larger forum? What is there to hide regarding one's meme use? Common meme-hosting sites are not actually private forums; posts are considered public and have been viewed by law enforcement and potential employers. Therefore, one must question, *What, from a meme standpoint, is going on in in the shadows of these secret Facebook groups?*

Continuing the speech, the AI claims that people begin "leaking whatever 'truth' suites them into the growing cesspool of society at large. The different cardinal truths neither clash nor mesh, no one is invalidated but nobody is right." This line has been a chillingly accurate predictor of meme use in the twenty-first century. Subreddits and other online forums known for meme use have devolved into echo chambers, where voices of dissent have been censored. We've also explored how memes are a poor medium when it comes to building cross-cultural bridges. A common feature of Reddit is the downvote button, where users can vote to send a post to the bottom of the feed, effectively exercising a censorship-by-group mechanic. This does nothing but encourage hive-mind behavior (akin to what we see in secret Facebook groups, where there exists an ideological gatekeeper). Since voices of dissent are effectively censored, it stands to reason that those who disagree with the group—or who attempt to engage the hive mind in dialog that runs counter to their dogma—are left to search for a group that welcomes their ideals. Because of this, meme groups rarely engage with groups of other opinions—and if they do, the conversations aren't typically productive.

The AI lectures Raiden, "Not even natural selection can take place here. The world is being engulfed in truth," a *truth* that is merely the product of confirmation bias. Natural selection depends on a binary system, one in which an organism either survives to pass on its genetic sequence or does not. In the *memeosphere*, memes are created and shared before fading into obscurity. But one must be careful not to confuse *fading into obscurity* with *death*. In this world, posts, no matter how short-lived or irrelevant they may be, are often permanent additions to the digital landscape. Rather than dying, this fading into obscurity leaves the common meme open to a resurrection, often multiple times. It isn't rare to log on to r/DankMemes to see a post complaining about recycled content. If meme posters are focused on recycling old content, then how could they possibly be trying to improve their craft?

Further, the AI laments the constant flow of pointless postings by the general public, saying its goal was to "wade through the sea of garbage you people produce." It is alarming how many confirmed meme archetypes have been entered into KnowYourMeme.com. KnowYourMeme serves as the internet's de facto database for all things *meme* (though as established before, the term *database* isn't quite appropriate). Presently, the cataloging site has a little over 4,800 known memes documented. This is merely a drop in the ocean when viewed in the context of the billions of users that common meme-hosting websites such as Facebook, Twitter, and Reddit are known to have. The users of these sites flood the internet with low-effort memes (a literal and unironic flair on the r/DankMemes subreddit). Of the millions of memes posted daily, very few will make it to the front page of Reddit, "go viral" on Facebook, or reach any other quantifiable definition of success. The rest are, statistically speaking, doomed to a life of obscurity with the chance of occasional recycling.

The AI presses Raiden further, telling him that "a single person has the potential to ruin the world. And the age of digitized communication has given more power to the individual." Thanks to easily spreadable and digestible memes, this is a true statement. As mentioned before, memes have a low barrier to entry, and hence any individual, regardless of artistic or communicative skill, can create and proliferate their ideals in meme form. On the off chance a particular meme becomes viral, the meme has the potential to influence millions of viewers, whether the meme is benign or radicalized (though, thankfully, as we explored in the activism chapter, the medium rarely encourages real-world activity).

In the final relevant part of the speech (as it pertains to this book), the AI tells Raiden that "anything can be quantified nowadays." This is an especially pungent reality in this post-2016 Cambridge Analytica world. The Weng and Bauckhage studies cited earlier lend credibility to this claim of memes serving as quantifiable data points. Common meme-hosting websites have all scrupulously designed algorithms to scrape every last bit of data regarding what pages and posts a user likes, dislikes, and shares. This data can then be used to design a propaganda campaign specifically tailored to that user, to a group of users, or to society as a whole. Such manipulation was indeed the goal of the AI in *Metal Gear Solid 2*, which sought to control the flow of information, public discourse, and subsequently, human thought and behavior. Juxtapose that mission statement with the Cambridge Analytica scandals regarding Brexit and the 2016 US presidential election. A fictional video game released in 2001 turned out to be incredibly prescient in describing our societal meme-based norms. Memes are not our friends.

ARE MEMES AN EFFECTIVE COPING MECHANISM?

MEMES HAVE BEEN STUDIED for their use as a coping mechanism for stress. In their study of the use of memes in Puerto Rico during the lockdown caused by COVID-19, Flecha-Ortiz et al. observed that memes were used to express differing interpretations of the same event. A common theme present in the medium was the use of humor to cope with the stressful situation. It is not the stance of this book to say that stress relief isn't important; however, it *is* the stance of this book to say that action must follow stress relief and, in this respect, memes are inherently flawed.

Flecha-Ortz et al. observed that memes used emotion, in this case, primarily humor, to guide their users. While their use in the context of COVID-19 was as a well-intended stress reliever, we as a society need to be aware that memes can be used by nefarious actors to guide a desired viewpoint. The researchers also made a point to mention that variation in the novelty of facts displayed in memes can impact their interpretations. This isn't to say that memes outright lie or are *fake news*, per se, but it could mean that memes use old and outdated information to intentionally manipulate the viewers. As we discussed before, memes are vetted for ideology instead of accuracy.

Jessica Myrick, Robin Nabi, and Nicholas Eng (a trio of communications professors) performed a different study regarding the use of memes as coping mechanisms during the COVID-19 pandemic.[1] One of their findings was that, among other things, meme viewing was associated with lower levels of information processing. This is not a positive outcome for the communication medium. Users having less intellectual bandwidth may help in the short term to deal with stress in their day-to-day life but that absolutely is not going to help improve their situation over the long term, and hence is not an effective coping mechanism.

Myrick, Nabi, and Eng measure several different variables regarding the effectiveness of the medium to help individuals cope with stress. One popular format of the medium is the use of animals, specifically small and young animals (newborn puppies, kittens, etc.) since they are widely perceived as cute. However, in this study, memes with cute animals were associated with lower levels of both information processing and coping efficacy. For a communication medium that many say helps them cope with the demands of everyday living, memes have been strikingly ineffective in that respect. Evidence suggests that memes actually make people less perceptive of stimuli. This strategy is akin to removing the batteries from the smoke detector; doing so focuses on the wrong stimuli (the peskiness of the alarm chirping versus the danger to life and limb).

The lower levels of information processing discussed in the last paragraph are troubling. Myrick, Nabi, and Eng found an association between information processing and coping efficacy. So, if memes

---

1   Myrick, Jessica Gall, et al. "Consuming Memes during the COVID Pandemic: Effects of Memes and Meme Type on COVID-Related Stress and Coping Efficacy." *Psychology of Popular Media*, 18 Oct. 2021, 10.1037/ppm0000371. Accessed 27 Nov. 2021.

lower someone's ability to process information, they also lower the users' ability to cope with stress. Upon doing research for this book, many people have stated that they prefer to peruse memes after a long day at work, implying that memes are used for some sort of stress relief (consciously or otherwise). Given these findings, memes do not help people cope with stressors in their lives; they merely numb the mind to said stressors. The effect of memes is similar to alcohol and illegal drug use, which no expert would recommend for coping with stress.

Humor has long been hailed as another means of coping with stressors in one's life. Admittedly, the appeal is there. However, Myrich, Nabi, and Eng found that humorous internet memes were not associated with coping efficacy in their participants. This is a devastating blow to the pro-meme community. Often, the biggest pitch meme apologists make is the positive use of humor in the medium; however, this study shows that humor is not an effective tool for this desired outcome.

Myrich, Nabi, and Eng found that captions of memes containing information regarding COVID-19 were found to relieve stress in their participants. But, as we have discussed in the earlier chapters, memes are often fraught with inaccuracies and outright lies due to the illusory truth effect, which is abundant in the medium. So, captions containing information about the virus could easily contain false information. We must ask ourselves, *What are we trying to accomplish when the medium's most effective means of stress relief cannot even be confirmed as truthful?* Furthermore, why are we seeking out memes instead of making public service announcements for such valuable information?

The Myrick, Nabi, and Eng study ends with the admission that viewing three memes was sufficient to provide the desired stress relief effects. Given the addictive nature of memes, the prolonged length of visits on these websites and the echo chamber that memes are known to form leads us to believe that the average meme user consumes far more than three memes per day. This sort of metaphorical overdose clearly cannot be healthy for the user (and therefore, for society as a whole, given how widespread the meme-use problem is). Since memes are literally habit-forming (as Cannizzaro noted), recommending meme intake of any amount is playing a dangerous game and is a slippery slope.

Memes featuring alcohol use as a coping mechanism have become alarmingly common. Historian and folklorist James Seaver discusses the trend of wine-related memes involving American mothers.[2] The memes studied routinely reinforced the practice of drinking vast amounts of wine (often to the point of being habit-forming); however, memes revolving around criticizing the practice were few and far between. Neither excessive wine consumption nor excessive *memeing* about alcoholism are effective coping mechanisms.

Seaver argues that the growing trend of alcoholism in the US among married, middle/upper-class, White mothers between the ages of twenty-five and thirty-four is correlated to the wine-related memes. Once again, the concept of *you are what you meme* absolutely rings true in this context. That is to say, if one posts several memes normalizing their growing alcoholism, then it is likely that they themselves have a growing problem with alcohol. But this also dovetails with the addictive nature of the communication medium as a whole. Given that Seaver references a secret Facebook group, it is impossible to tell how many memes on the topic of female alcoholism have been posted. While this is speculation, it is a safe guess to say that the number of memes about alcoholism in this secret group surpasses the three-meme limit proposed by the Myrick study.

Seaver noticed an important trend with the memes examined in his article—they all trivialized a serious problem. As we have seen, memes tend to use humor as a way of veiling the seriousness of the content displayed within. Seaver discusses how female alcoholism

---

2   Seaver, James. • *Articles* • *Mommy Juice: Internet Memes and the Dark Humor of Wine Consumption among American Mothers.* 2020.

has been on the rise in the United States since 1999 (again, 1999 serves as a good "pre-meme" proxy), The point isn't that the wine memes have caused these women to drink excessively, but memes *did* make the behavior culturally okay, even if it tacitly so.

Memes about addictive substances are especially dangerous because, as discussed earlier, memes as a communication medium reinforce cultural norms and are used to police their members, without creating any context or allowing for dissent. Memes in their own right are habit-forming, as discussed by Cannizzaro. An addictive medium spewing a playful message about a dangerously addictive substance is a combination that should not be taken lightly. This is not what an effective coping mechanism looks like.

Seaver also goes on to discuss how the wine-related memes tended to "pathologize" motherhood, with Seaver musing that the memes made motherhood out to be something that others needed to self-medicate from. This book is not a parenting book and will not comment on the difficulties of parenthood. Here, we can simply observe that memes failed in their task of being an effective coping mechanism because they did nothing to solve the grander societal issues that enabled the rise in wine consumption. Specifically, Seaver notes, the wine memes did not address critical issues such as income disparity, support from partners, affordable childcare, paid family leave, and more. Since these memes did not solve (or even *mention*) these problems, and since memes are known to be habit-forming, we can confidently say that memes are not an effective coping mechanism, nor do they help people deal with their stressors.

Seaver aptly points out how misery loves company, more so when that company uses the type of humor entrenched in memes to reinforce norms. Worse yet, memes created the type of environment where refusing to affirm the behavior condoned within a meme (in this case, wine-based alcoholism) can ostracize a member from the existing group. We have seen this behavior in our chapter on extremism. Memes do not create communities of support, which in this context seems to be sorely needed. Among the echo chambers formed by memes, we can find silent members, wishing they did not find themselves in isolation.

Other studies have demonstrated the failure of memes as a means of coping. Kertcher and Turin discussed the memes used during COVID-19 lockdowns in Israel. One of the biggest supposed selling points of memes is their ability to use humor as a coping mechanism. However, Kertcher and Turin found that, in their 436 memes sampled during that time period, less than 10 percent of memes used humor in discussing political turmoil in the nation (in fact, memes about political turmoil were largely absent). Therefore, when an effective coping mechanism was needed the most, memes failed to live up to their major selling point. This goes to show that memes are an ineffective coping mechanism, despite claims to the contrary.

Drury explores the use of memes as a stress coping mechanism by emergency medical technicians (EMT) and paramedics. The paper started off by providing a lot of necessary background information regarding the day-to-day life of an EMT (it's important to note that this book concedes that EMS personnel have a difficult job). However, as the paper examined memes regarding the profession (as uncovered in the subreddit r/EMS), the author noticed some general trends regarding meme usage. One common theme in the medium was the use of mockery and disdain toward patients and "bullshit calls." One meme mocked a patient who had tooth pain and needed an ambulance ride, encouraging utter contempt toward the patient in question. Thankfully, a commenter gave additional context that tooth pain is a sign of a rare but usually fatal type of heart attack. Without that context (pointedly, context *not* given by the meme), the audience would likely have continued their disgust with the patient. In this example, memes acted not as a coping mechanism but instead as a means of quite literally *putting people's lives in danger*.

Drury continued that many of the memes observed relating

to the EMT/paramedic profession were predominately negative, disparaging the healthcare system and patients alike. For example, there was a common theme that taxis and ambulances are equivalent and patients abuse the system. These memes are not helping their audience cope with or improve their situation regarding the difficulty of their job. Memes are not doing anything to improve the relationship between patient and caregiver (one could argue, through the evidence presented, that memes are making that *worse*), nor are memes helping to repair what many call a broken healthcare system.

Drury ends the article with two admissions; memes do not directly address the issue of EMT/paramedic burnout, and it is impossible to tell how memes impacted individuals specifically. We have already explored how memes, despite being hailed as supposed stress relievers, do nothing to actually relieve the stress of job difficulties or help individuals cope long term. Also, without the knowledge of how memes impact the individual, the claims of memes being an effective coping mechanism are, therefore, baseless. Making such grandiose claims without any backing is not only unethical but can be downright dangerous, given the mental health of those tasked with administering emergency care. Memes are not an effective coping mechanism. Rather, memes make society sicker.

SUMMARY AND CONCLUSION

WHAT THIS BOOK SUGGESTS is not legislative change but rather a social change. After all, passing new laws is unlikely to work long-term as people can finagle around the letter of the law. Passing new legislation is also undesirable due to the unintended consequences such a law (or set of laws) would create. Further, advocating the legislative branch of the government to get involved would be a gross misuse of their time, assuming such a measure would gain any support in the first place. There is also the risk of our legislators crafting a law (or set of laws) that acts as little more than a power grab, in an attempt to curb *memeing*, as we saw in Belarus and Singapore. Therefore, a much more effective long-term solution is to change the hearts and minds of those who hold *memeing* dear. They need our embrace, not our regulation.

In the background chapter, we laid the foundation of what memes are, viewing them especially through the lens of the Richard Dawkins. Using that as our framework, it was important to identify why memes have always been a flawed communication medium. Many of the pro-meme papers that Big Academia produces frequently cite Dawkins's work without fully realizing (or worse yet, blatantly ignoring) the flaws of meme theory. With the admittedly shaky foundation that memes are built on, it wasn't difficult to start methodically pulling apart the medium.

In the creativity chapter, we examined how memes are not a creative artistic medium, in either the technical aspects of making art (from a graphic design standpoint) or from a piracy standpoint. The literal etymology of the word shows that the medium was never intended to be one where a so-called *creator* can thrive. Piracy and plagiarism run rampant through the medium, on a scale that is difficult to replicate in other artistic mediums. Memes do not come with a works cited page to properly credit the author, nor do they come with a certificate of authenticity or a serial number. We have demanded lot traceability with commodities such as diamonds to ensure they weren't involved in nefarious dealings, yet we have made no such demands for memes, unfortunately.

In the context chapter, we explored how the medium works, or more pointedly how it *doesn't* work to build a greater understanding. The medium relies on short, quippy, two-line structures to make a point. The medium assumes that the audience is inherently *in on it* and does nothing to build any context for those who aren't. Memes cannot survive on their own.

When we discussed civil discourse as it relates to memes, we explored the point their proponents claim that *memes start discussions.* However, we unearthed that the discussions that memes start are typically rife with confirmation bias. Memes cause an ideological echo chamber to form. We also discussed how meme users expect to be able to change the meaning of language itself, yet have been shown to be unsuccessful at doing so through their medium of choice. It is interesting to note that memes are inherently reliant upon written language to make their points (however vapid those points may ultimately be). We also explored that the medium displays varying interpretations over time, ultimately converging them to form a singular hive-mind opinion (if you don't believe me, ask any Redditor; they'll both participate in and bemoan the hive mind at the same time). Memes flatly do not create engaging civil discourse; they destroy it.

In the education chapter, we looked at the use of memes in classrooms at varying grade levels. At the very top, we examined veterinary and pharmacy students—people who will one day be tasked with the health and well-being of our society—let down by the shortcomings of memes in the classroom. The use of memes in courses of engineering students in India, a blossoming economy that produces many top scientists and engineers, failed to build new concepts in the mind of students. The same held true of high school math students in Italy. Across international borders, languages, and cultures, memes have established themselves as a global problem. Memes are simply not an effective teaching tool, given the evidence presented in this book. Write your local elected officials and school board, and implore them to cease the use of internet memes in the classroom. The future leaders of our society deserve better than to be subjected to memes. But this book is not opposed to technology as a supplemental learning tool. One example of a great website to do this is KhanAcademy.com, where there are instructor-lead videos on a seemingly endless list of topics. The internet is a wide-open sea of educational opportunities; let's not settle for memes.

Memes have been shown to be addictive. We briefly explored the scientific backing of the addiction feedback loop. We discussed how memes benefit only the Silicon Valley Elite, to the detriment of our wider economy and our communities. Memes were designed to addict, as Richard Dawkins himself stated, "When you plant a fertile meme in my mind, you literally parasitize my brain." The comparison to a parasite is a chillingly accurate one. *Memes are not your friends.*

Furthermore, we discussed the role of humor used in memes. Humor is an overused delivery vehicle to convey the message of a meme. As we saw in the education chapter, humorous memes often cause the audience to make light of the often serious subject matter. Worse yet, humor within the medium is used to veil harmful and hateful content so the message can be dismissed as "just a joke." This is a dangerous social game to play because it can (and often does) result in the segregation of an "us" verses a "them," as we examined in 4chan boards. One must be very skeptical of what is observed from a humor standpoint in memes, as a humorous meme often hides a devious message.

In the journalism chapter, we unpacked how memes are not a reliable source of news, and journalists should not rely on them. The use of memes as journalistic sources violates several articles of the Code of Ethics created by the Society of Professional Journalists. Memes are a reactionary medium and thus are not even a timely source of news. This is because the event needs to happen before memes can be created, and memes cannot truly capture a story as it unfolds. Put another way, memes can only look backwards. Memes, when used as the focal point of reporting, lead to potentially widespread exploitation of volunteer contributors (after all, very few people are paid to meme).

Memes, as we unpacked in another chapter, do not encourage real-world activism either. Whether it was in constant monitoring regimes, such as Singapore and Belarus, or in more liberal countries like the United States and Estonia, memes have shown to be an ineffectual tool in encouraging people to gather in real life for the causes they champion. Memes are known to encourage *slacktivism*— the practice of performative support for a cause on the internet without any real-world action. Memes serve only to signal to others; when the signaling is complete, so is the meme posters' involvement in their supposed cause. As we saw with the Leonard Peltier case study, memes that had a very specific call to action had a better response than those that didn't; however, this tactic is woefully underrepresented in the medium.

Memes are a highly disposable communication medium, as we explored in the relevancy chapter. The groundbreaking Weng study quantified that memes have an alarmingly short shelf life, often losing their peak appeal within *hours*. Memes are often held up by their proponents as pieces of digital folklore akin to surviving for generations. However, the Weng study quantifiably shows that this is not the case. After their brief window of opportunity for internet fame, memes die an often-forgettable death, as the meme cycle washes anew frequently (and in the case of Reddit, the bastion of memes, the front page changes every twenty-four hours).

Memes as a medium are very conducive to those spouting extremist views. Memes have a desensitizing effect upon their audience, as noted by Hecker's study. The desensitizing effect impacts the user's sense of self, as mentioned by Yus's study, meaning that, in a very real way, *you are what you meme* (contrary to the common meme lover's defense of *it's just a joke*). Desensitization and the illusory truth effect—memes can serve as effective tools to those looking to radicalize new members toward their hate groups. But, even in such extremist settings, memes remain a poor communicator of cross-cultural ideas. Memes were ineffectual at building bridges between two extremist groups with similar ideologies and, for this purpose, longer-form text posts were used to communicate across groups. Fortunately, as explored in the activism chapter, using memes doesn't correlate with real-world action most of the time. The pitfall of the medium serves as a *de facto* safety valve.

A common defense of the medium is that *memes are participatory*. In that chapter, we discuss that memes are not truly participatory for all. Instead, memes are more of a megaphone for one to broadcast their views at the expense of all others around them. To everyone else around the megaphone user, this is not participatory, since mutual consent is not necessary in the use of memes. Memes also do not truly encourage participation, because the barriers to entry of many common meme-hosting sites like Reddit can remain high (and one must have the qualifying amount of karma in order to post on certain subreddits).

Lastly, we looked at memes as a coping mechanism for stress, as that is a popular claim/practice of many users (who, for example, scroll through memes after a long day at work). Memes are ineffective at providing a solution to one's long-term problems or daily stressors. Furthermore, memes can make dangerous behaviors like alcoholism tacitly acceptable, which only serves to make daily problems worse. Memes also breed contempt toward one's daily work, as we saw in the case of the emergency medical personnel. Memes do not help people cope with the demands of everyday life.

This book has been heavy on scholarly research and peer-reviewed scientific articles for a reason; the first priority of this book has been to demonstrate that the problem of memes actually exists—and that it is pressing. Society has been in denial of the pitfalls about *memeing* for too long, to the point of vilifying those with openly anti-meme viewpoints. Put another way, before long-term change can occur, we, as a society (and, as the evidence presented can attest, societies around the world), must first acknowledge that memes have reached a crisis point.

Drackett touched upon another valid point, one which is the crux that inspired this entire book—the inherent pro-meme bias of academia. Their analysis found that academia has a staunch bias to only funding research and publishing the benefits of meme-based humor, while very little attention is paid to the pitfalls of the medium. Furthermore, Big Academia has softened their definitions of terms such as *participation, mobilization,* and *activism* to support whatever conclusion they were hoping to achieve. **That *is not* how science works**. Their bias to sweep the negative aspects of memes under the rug has not been contained to merely the ivy-coated walls of academia. As stated in the introduction, society itself has adopted a pro-meme stance, while vastly ignoring or downplaying the inherent flaws of the communication medium. We've also touched upon how the Silicon Valley Elites are incentivized to not do anything about the problem either.

However, that isn't to say that we cannot try to offer solutions. While up to this point this book has been based in evidentiary studies, the closing recommendations will be both speculative and based in opinion (although, these speculations and opinions are based on the evidence presented in this book). After all, it is of a limited use to merely point out a problem without offering a solution; such a behavior takes away the onus to act.

One solution is to not laugh or positively reinforce the memes used in communication. It is imperative to deny those memes their sought-after social validation. It is also equally important to not use the mechanisms of approval common on many meme-hosting platform (likes on Facebook, upvotes on Reddit, etc.). When memes lose the strategic edge of humor and the advantage of social signaling, they will lose much of their power. And, by all means, *do not spread memes*. The proliferation of memes has been a front and center justification for their discussion, dating as far back as 1976, when Richard Dawkins discussed their spread in *The Selfish Gene*. While the temptation may exist, do not engage in counter-*memeing*; this is rarely an effective strategy (as we saw with the Kony2012 craze). The only good meme is a deleted meme.

But refusing to react is not enough. Silence can be taken as complicit agreement with the use of memes. Collectively, we must stand and demand those who use memes express their ideas in other vehicles—namely a *longer-form* medium. Demand background information. Push for more context. Ask the poster to fully articulate their stance on whatever issue they claim to embrace, and make them use complete sentences and fully formed arguments instead of two-line quips. Take the interaction seriously, and do not allow for a humorous distraction. Demand evidence for their claims, because as Richard Dawkins himself said, "Nothing is more lethal for certain kinds of meme than a tendency to look for evidence." Most of all, be public in doing so; break up the echo chamber.

While we are at it, we need to send the collective message that meme use by adults is not okay. Upon high school graduation, society expects its citizens to be able to properly communicate with and actively build up the rest of society. Therefore, let's make a motion that *memeing* by those over the age of eighteen be discontinued. On that same note, we should not only refuse to use memes in the classroom, but we also need to educate our youth as to the pitfalls of the common meme. Stemming this tide before they reach adulthood is a generational solution to the problem of memes.

For the meme hobbyist in your life, take the time to build a real-world connection with them. We've explored how memes do not build cultural bridges very well. We've also explored that meme usage tends to be a self-isolating hobby. Demonizing those who use memes will not solve the problem. This book serves to attack the medium itself and *not* the people who use them.

ACKNOWLEDGMENTS

I WOULD LIKE to acknowledge and thank Dr. Steven P. Miller for all of his assistance during the developmental edit of this book. His feedback was vital.

Akram, Umair, Jason G. Ellis, Glhenda Cau, Frayer Hershaw, Ashlieen Rajenthran, Mollie Lowe, Carissa Trommelen, and Jennifer Drabble. "Eye Tracking and Attentional Bias for Depressive Internet Memes in Depression." *Experimental Brain Research* 239, no. 2 (December 17, 2020): 575–81. https://doi.org/10.1007/s00221-020-06001-8.

Aldric, Anna. "Average SAT Scores over Time: 1972 - 2019." Prepscholar.com, 2021. https://blog.prepscholar.com/average-sat-scores-over-time.

Anti-Defamation League. "Pepe the Frog." Anti-Defamation League, 2016. https://www.adl.org/education/references/hate-symbols/pepe-the-frog.

AriesStark. "Malicious Advice Mallard." Imgflip, 2015. https://imgflip.com/i/fzkl1.

Augustyn, Adam. "Finland - Nordic Cooperation | Britannica." www.britannica.com, 2010. https://www.britannica.com/place/Finland/Nordic-cooperation.

Bauckhage, Christian. "Insights into Internet Memes." University of Bonn, 2011.

Bebic, Domagoj, and Marija Volarevic. "Do Not Mess with a Meme: The Use of Viral Content in Communicating Politics." *Communication & Society* 313, no. 3 (2018).

Bede, Isabelle. "Journalism Embeds Social Media Language: The Use of Internet Memes in Political News." Masters Thesis, 2019.

Begg, Ian Maynard, Ann Anas, and Suzanne Farinacci. "Dissociation of Processes in Belief: Source Recollection, Statement Familiarity, and the Illusion of Truth." *Journal of Experimental Psychology: General* 121, no. 4 (1992): 446–58. https://doi.org/10.1037/0096-3445.121.4.446.

Bini, Giulia, and Ornella Robutti. "THINKING inside the POST: INVESTIGATING the DIDACTICAL USE of MATHEMATICAL INTERNET MEMES." *International Group for the Psychology of Mathematics Education*, 2019. http://hdl.handle.net/2318/1698304.

Brown, Joshua D. "What Do You Meme, Professor? An Experiment Using 'Memes' in Pharmacy Education." *Pharmacy* 8, no. 4 (October 29, 2020): 202. https://doi.org/10.3390/pharmacy8040202.

Bumiller, Elisabeth. "We Have Met the Enemy and He Is PowerPoint." *The New York Times*, April 26, 2010, sec. World. https://www.nytimes.com/2010/04/27/world/27powerpoint.html.

Cannizzaro, Sara. "Internet Memes as Internet Signs: A Semiotic View of Digital Culture." *Sign Systems Studies* 44, no. 4 (December 31, 2016): 562–86. https://doi.org/10.12697/sss.2016.44.4.05.

Crawford, Blyth, Florence Keen, and Guillermo Suarez-Tangil. "Memes, Radicalisation, and the Promotion of Violence on Chan Sites." *The International AAAI Conference on Web and Social Media (ICWSM)*, June 7, 2021. https://kclpure.kcl. ac.uk/portal/en/publications/memes-radicalisation-and-the-promotion-of-violence-on-chan-sites(ec3ef161-783e-4403-a09d-4858f42647df).html.

Denisova, Anastasia. "Political Memes as Tools of Dissent and Alternative Digital Activism in the Russian-Language Twitter." 2016.

Drakett, Jessica, Bridgette Rickett, Katy Day, and Kate Milnes. "Old Jokes, New Media – Online Sexism and Constructions of Gender in Internet Memes." *Feminism & Psychology* 28, no. 1 (February 2018): 109–27. https://doi.org/10.1177/0959353517727560.

Drury, Caroline. "The Function of Internet Memes in Helping EMS Providers Cope with Stress and Burnout." 2019.

Flecha Ortiz, José A, Maria A Santos Corrada, Evelyn Lopez, and Virgin Dones. "Analysis of the Use of Memes as an Exponent of Collective Coping during COVID-19 in Puerto Rico." *Media International Australia* 178, no. 1 (October 24, 2020): 168–81. https://doi.org/10.1177/1329878x20966379.

"Funny Animal Memes." Slapwank.com, 2017. https://slapwank. com/wp-content/uploads/2017/03/Funny-Animal-Memes-3. jpg.

Gilbert, Nestor. "48 Reddit Statistics You Must Read: 2020/2021 Data Analysis & Market Share." Financesonline.com, August 13, 2019. https://financesonline.com/reddit-statistics/#:~:text=1%20Almost%20half%20%2841.2%25%29%20 of%20Reddit%E2%80%99s%20visitors%20are.

Giovanni. "The Best 2016 Political Memes." Urban Myths, March 5, 2016. http://www.urbanmyths.com/urban-myths/politics/the-best-2016-political-memes/.

Godwin. "Meme, Counter-Meme." Wired, 1994. http://www.wired. com/wired/archive/2.10/ godwin.if_pr.html.

Gramlich, John. "10 Facts about Americans and Facebook." Pew Research Center, June 1, 2021. https://www.pewresearch.org/ fact-tank/2021/06/01/facts-about-americans-and-facebook/.

Hakoköngäs, Eemeli, Otto Halmesvaara, and Inari Sakki. "Persuasion through Bitter Humor: Multimodal Discourse Analysis of Rhetoric in Internet Memes of Two Far-Right Groups in Finland." Social Media + Society 6, no. 2 (April 2020): 205630512092157. https://doi.org/10.1177/2056305120921575.

Hardwick, Joshua. "Top 100 Most Visited Websites by Search Traffic (as of 2021)." SEO Blog by Ahrefs, 2021. https://ahrefs.com/blog/ most-visited-websites/.

Hecker, Rabea. "Are You Serious It Is Just a Joke? The Influence of Internet Memes on the Perception and Interpretation of Online Communication in Social Media." 2020.

Hernandez-Cuevas, Eva Marie. "The Pertinence of Studying Memes in the Social Sciences." Kansas City University, 2021. https:// www.researchgate.net/publication/352258835.

"High Expectations Asian Father." Imgflip, 2011. https://imgflip. com/i/m2.

Huntington, Hiedi. "Subversive Memes: Internet Memes as a Form of Visual Rhetoric." Selected Papers of Internet Research, 2013.

"Image Tagged in Republican,Hate,Congress." Imgflip, 2020. https:// imgflip.com/i/2vhv1w.

Karatzogianni, Athina, Galina Miazhevich, and Anastasia Denisova. "A Comparative Cyberconflict Analysis of Digital Activism across Post-Soviet Countries." Comparative Sociology 16, no. 1 (February 13, 2017): 102–26. https://doi.org/10.1163/15691330-12341415.

Kertcher, Chen, and Ornat Turin. "'Siege Mentality' Reaction to the Pandemic: Israeli Memes during Covid-19." *Postdigital Science and Education* 2, no. 3 (August 5, 2020): 581–87. https://doi. org/10.1007/s42438-020-00175-8.

Kien, Grant. "Media Memes and Prosumerist Ethics." *Cultural Studies Critical Methodologies* 13, no. 6 (November 12, 2013): 554–61. https://doi.org/10.1177/1532708613503785.

Kojima, Hideo. "Metal Gear Solid 2: Sons of Liberty." Video Game. 2001. https://www.youtube.com/watch?v=eKl6WjfDqYA.

Lenhardt, Corinna. "'Free Peltier Now!' the Use of Internet Memes in American Indian Activism." *American Indian Culture and Research Journal* 40, no. 3 (January 1, 2016): 67–84. https://doi. org/10.17953/aicrj.40.3.lenhardt.

Loughlin, Wendy. "Meme Effects: Assistant Professor Rebecca Ortiz Provides Insight on a Powerful Form of Communication." Newhouse School at Syracuse University, October 28, 2020. https://newhouse.syr.edu/news/meme-effects-assistant-professor-rebecca-ortiz-provides-insight-on-a-powerful-form-of-communication/.

"Malicious Advice Mallard." Imgflip. Accessed February 6, 2022. https://imgflip.com/i/fzkl1.

María Hernández-Cuevas, Eva. "THE PERTINENCE of STUDYING MEMES in the SOCIAL SCIENCES Meme Sharing 'Culture' and Psychological Well-Being from a Neuroscientific Framework View Project Mental Health and Internet Memes View Project." Revista Ingenios, 2021.

Massanari, Adrienne. "Participatory Culture, Community, and Play: Learning from Reddit." New York: Peter Lang, 2015.

"Memes." Anti-Democrat Stickers. Accessed 2021. http:// antidemstickers.com/memes/.

Myrick, Jessica Gall, Robin L. Nabi, and Nicholas J. Eng. "Consuming Memes during the COVID Pandemic: Effects of Memes and Meme Type on COVID-Related Stress and Coping Efficacy." *Psychology of Popular Media*, October 18, 2021. https://doi.org/10.1037/ppm0000371.

National Assessment of Educational Progress. "Reading and Mathematics Score Trends," 2016. https://nces.ed.gov/programs/coe/pdf/coe_cnj.pdf.

National Coalition of Core Arts Standards. *National Core Arts Standards-Visual Arts at a Galance*. Dover, Delaware: National Core Arts Standards, 2014.

Newcomb, Alyssa. "Who Is Policing Facebook's Secret Groups?" NBC News, 2017. https://www.nbcnews.com/tech/social-media/who-s-policing-facebook-s-secret-groups-n729856.

Newman, Eryn J., Maryanne Garry, Daniel M. Bernstein, Justin Kantner, and D. Stephen Lindsay. "Nonprobative Photographs (or Words) Inflate Truthiness." *Psychonomic Bulletin & Review* 19, no. 5 (August 7, 2012): 969–74. https://doi.org/10.3758/s13423-012-0292-0.

Noyes, Jillian. "Election 2016: The Year of the Meme." The Odyssey Online, November 7, 2016. https://www.theodysseyonline.com/election-2016-year-meme.

Patel, Ronak. "First World Problems:' a Fair Use Analysis of Internet Memes." escholarship.org, 2013. https://escholarship.org/uc/item/96h003jt.

Pew Research Center. "Social Media Fact Sheet." Pew Research Center: Internet, Science & Tech. Pew Research Center, April 7, 2021. https://www.pewresearch.org/internet/fact-sheet/social-media/.

PhantasmGear. "Pin by Phantasmgear.com on Overwatch | Overly Attached Girlfriend, Real Estate Memes, Girlfriend Meme." Pinterest. Accessed March 21, 2022. https://www.pinterest.com/pin/845691636250276337/.

Preez, Amanda du, and Elanie Lombard. "The Role of Memes in the Construction of Facebook Personae." *Communicatio* 40, no. 3 (July 3, 2014): 253–70. https://doi.org/10.1080/02500167.2014.938671.

Reber, Rolf, and Norbert Schwarz. "Effects of Perceptual Fluency on Judgments of Truth." *Consciousness and Cognition* 8, no. 3 (September 1999): 338–42. https://doi.org/10.1006/ccog.1999.0386.

Reddy, Rishabh, Rishabh Singh, Vidhi Kapoor, and Prathamesh P Churi. "Joy of Learning through Internet Memes." *International Journal of Engineering Pedagogy (IJEP)* 10, no. 5 (October 15, 2020): 116. https://doi.org/10.3991/ijep.v10i5.15211.

Richard Dawkins. *The Selfish Gene.* Oxford: Oxford University Press, 1976.

Root Kustritz, Margaret V. "Effect of Differing PowerPoint Slide Design on Multiple-Choice Test Scores for Assessment of Knowledge and Retention in a Theriogenology Course." *Journal of Veterinary Medical Education* 41, no. 3 (September 2014): 311–17. https://doi.org/10.3138/jvme.0114-004r.

Seaver, James. "♦ Articles • Mommy Juice: Internet Memes and the Dark Humor of Wine Consumption among American Mothers," 2020.

Sen, Ari, and Brandy Zadrozny. "QAnon Groups Have Millions of Members on Facebook, Documents Show." NBC News, 2020. https://www.nbcnews.com/tech/tech-news/qanon-groups-have-millions-members-facebook-documents-show-n1236317.

"Shallow Rave: When Memes Go to War; KONY 2012." Shallow Rave, March 10, 2012. https://shallowrave.blogspot.com/2012/03/when-memes-go-to-war-kony-2012.html.

smooveb. "SAVAGE Political Memes 3." www.ebaumsworld.com. Accessed 2021. https://www.ebaumsworld.com/pictures/savage-political-memes-3/85159074/.

Society of Professional Journalists (SPJ). *Code of Ethics*, 2014. https://www.spj.org/ethicscode.asp.

Soh, Wee Yang. "Digital Protest in Singapore: The Pragmatics of Political Internet Memes." Media, Culture and Society, 2020.

Stenovec, Timothy. "Myspace's Biggest Moments: Memories of a Fallen Social Network." HuffPost, June 29, 2011. https://www.huffpost.com/entry/myspace-history-timeline_n_887059?slideshow=true#gallery/5bb385dce4b0fa920b9b4ab1/18.

Swinyard, Holly. "Pepe the Frog Creator Wins $15,000 Settlement against Infowars." the Guardian. The Guardian, June 13, 2019. https://www.theguardian.com/books/2019/jun/13/pepe-the-frog-creator-wins-15000-settlement-against-infowars.

Takeuchi, Hikaru, Yasuyuki Taki, Kohei Asano, Michiko Asano, Yuko Sassa, Susumu Yokota, Yuka Kotozaki, Rui Nouchi, and Ryuta Kawashima. "Impact of Frequency of Internet Use on Development of Brain Structures and Verbal Intelligence: Longitudinal Analyses." *Human Brain Mapping* 39, no. 11 (June 28, 2018): 4471–79. https://doi.org/10.1002/hbm.24286.

UC Santa Barbara. "Voter Turnout in Presidential Elections | the American Presidency Project." www.presidency.ucsb.edu, 2021. https://www.presidency.ucsb.edu/statistics/data/voter-turnout-in-presidential-elections.

"Vaccines: Last Week Tonight with John Oliver (HBO)." www.youtube.com. John Oliver, June 2017. https://www.youtube.com/watch?v=7VG_s2PCH_c&t=988s.

Weng, L., A. Flammini, A. Vespignani, and F. Menczer. "Competition among Memes in a World with Limited Attention." *Scientific Reports* 2, no. 1 (March 29, 2012). https://doi.org/10.1038/srep00335.

Yus, Francisco. "Identity-Related Issues in Meme Communication." *Internet Pragmatics* 1, no. 1 (May 28, 2018): 113–33. https://doi.org/10.1075/ip.00006.yus.

CPSIA information can be obtained
at www.ICGtesting.com
Printed in the USA
BVHW040900211122
652279BV00026B/70